Steinbeis-Edition

Werner G. Faix, Jens Mergenthaler
Rolf-Jürgen Ahlers, Michael Auer

InnovationQuality

The Value of the New

Impressum

Werner G. Faix, Jens Mergenthaler, Rolf-Jürgen Ahlers, Michael Auer
InnovationQuality. The Value of the New

1st edition, 2015 | Steinbeis-Edition, Stuttgart
ISBN 978-3-95663-062-0
First published in 2014 | German edition "InnovationsQualität"

Layout: Steinbeis-Edition
Cover picture: © Shutterstock.com / watchara
Print: Printsystem GmbH, Heimsheim
Translation: englishtalk GbR, Stuttgart

Steinbeis is an international service provider in entrepreneurial knowledge and technology transfer. The Steinbeis Transfer Network is made up of about 1,000 enterprises. Specialized in chosen areas, Steinbeis Enterprises' portfolio of services covers research and development; consulting and expert reports as well as training and employee development for every sector of technology and management. Steinbeis Enterprises are frequently based at research institutions, especially universities, which are constituting the Network's primary sources of expertise. The Steinbeis Network comprises around 6,000 experts committed to practical transfer between academia and industry. Founded in 1971, the Steinbeis Foundation is the umbrella organization of the Steinbeis Transfer Network. It is headquartered in Stuttgart, Germany.

180936-2015-08 | www.steinbeis-edition.de

List of contents

List of figures

List of tables

List of formulas

Foreword

Since as far back as the times of Heraclitus, it has been self-evident and nevertheless true, that nothing is more permanent and nothing is more inevitable than change. The rollercoasters of history, great and small – disasters, blessings, twists and turns, ruin, transition, truncation – the chronicles of time on this planet are full of such developments. Even supposedly "quieter times" are ultimately just phases filled with smaller ups and downs. Change is a constant, and, as such, an intrinsic part of every era and every major challenge. It is something that every being on this planet has always had to – and will always have to – come to terms with.

From the things people say and write these days, one could even think that modern society is confronted by a degree of change never witnessed by any generation before. One also has the impression that, for the first time in history, humankind is facing many unknowns, unlike anything ever previously encountered. Given the scale of political, social and cultural change, given the volume of disruptive invention and new insights, given the number of major natural and manmade disasters that humanity has experienced during its existence, this view of the world should at least be seen in relative terms. Every generation has its own challenges to face, and, as Konrad Lorenz also believed, one thing holds true for every generation: You have to see if you stand or fall.

Without a doubt, however, the rate of change has accelerated in recent decades and the world is (again) witnessing more instability and unpredictability. Without a doubt, people are now living (again) in times of change in which many things (in some areas, possibly all things) are unlike anything that previous generations encountered.

Our world of today seems to be always on the go and moving on to the next development. And as a result, businesses and whole economies have to strive and dare to make evolutionary or even revolutionary changes themselves. The ability and the willingness to innovate means being able to and wanting to allow new things to become a reality – a reality that creates value and thus also adds value. And it is this which dictates the fate of, not just an economy, but also business.

Innovation is sometimes a radical (re)modeling of the existing, an act Joseph A. Schumpeter described as the "process of creative destruction," – a process that creates something new by completely eradicating something already established. It was also Schumpeter who identified how economies and companies can safeguard their competitiveness in the face of change and build on it:

"Profit [...] is the premium put upon successful innovation".

The results of our work, which are presented in this book, have motivated us and given us the courage to define and outline a new term, with the aim of making evident the actual nature and magnitude of the reward for successful innovation. We name this quantitative variable, which in a manner of speaking determines the value of "the New":

InnovationQuality InQ

This book was written along classic lines, with several people capturing the insights they have gained into this area in writing. And like any other academic publication, the list of references gives works written by others – works that the authors have pored over to gain some of these insights. Despite this, the statements we make are actually only a fraction of the thoughts and discussions we had while working on this book. A much larger share of these thoughts – thoughts which inspired us to write this book in the first place, and thoughts that inspired us as we were actually writing this book – stem from feedback from all the people who work in, work with, and more importantly, work on the development of the Steinbeis Network. So it is these people whom we wish to thank now, at the beginning, for their thoughts, for their work, and especially even for their disapproval – because when a system is subject to alternating current, there is the necessary degree of electricity and thus also power.

Werner G. Faix *Jens Mergenthaler* *Rolf-Jürgen Ahlers* *Michael Auer*

Stuttgart, November 2014

Introduction

1 The issue

"Nothing endures but change". (Heraklit) The idea that nothing is more constant than change has been around for a long time. Despite this, the oftentimes quoted “new normal” points to a kind of dramatic change that is becoming more the rule, in short cycles, than an exception – something that might just happen “sometime or other.” Managers in every corner of the globe expect the world in general and business in particular to become different – strikingly more dynamic, less certain, more complex and structurally different. (IBM 2010: 15) Or, as the head of a US government agency summarized it in the IBM 2012 Global CEO Study: “There isn’t a single day I come into work when I know what will happen.” (IBM 2012: 12)

But what led to this “new normal”? What are the drivers of this constant, and at the same time, dramatic change? And how could or should one react to it?

1.1 The ordinariness of incessant and intense change

Leading the way as a driver of change is the market economy, which, in its inherent nature, significantly contributes to permanent and at times dramatic change. Essential ingredients of a market economy are capitalism and entrepreneurs. In The Theory of Economic Development, Joseph Schumpeter attributes all development tendencies to the existence of creative entrepreneurs whose actions are focused on not permitting economic equilibrium in the form of absolute competition.

> *Capitalism, then, is by nature a form or method of economic change and not only never is but never can be stationary. And this evolutionary character of the capitalist process is not merely due to the fact that economic life goes on in a social and natural environment which changes and by its change alters the data of economic action; this fact is important and these changes (wars, revolutions and so on) often condition industrial change, but they are not its prime movers. Nor is this evolutionary character due to a quasi-automatic increase in population and capital or to the vagaries of monetary systems of*

which exactly the same thing holds true. The fundamental impulse that sets and keeps the capitalist engine in motion comes from the new consumers' goods, the new methods of production or transportation, the new markets, the new forms of industrial organization that capitalist enterprise creates.

[...] So is the history of the productive apparatus of the iron and steel industry from the charcoal furnace to our own type of furnace, or the history of the apparatus of power production from the overshot water wheel to the modern power plant, or the history of transportation from the mail-coach to the airplane. The opening up of new markets, foreign or domestic, and the organizational development from the craft shop and factory to such concerns as U. S. Steel illustrate the same process of industrial mutation - if I may use that biological term - that incessantly revolutionizes the economic structure from within, incessantly destroying the old one, incessantly creating a new one. This process of Creative Destruction is the essential fact about capitalism. It is what capitalism consists in and what every capitalist concern has got to live in. (Schumpeter 1942/1975: 82-83)

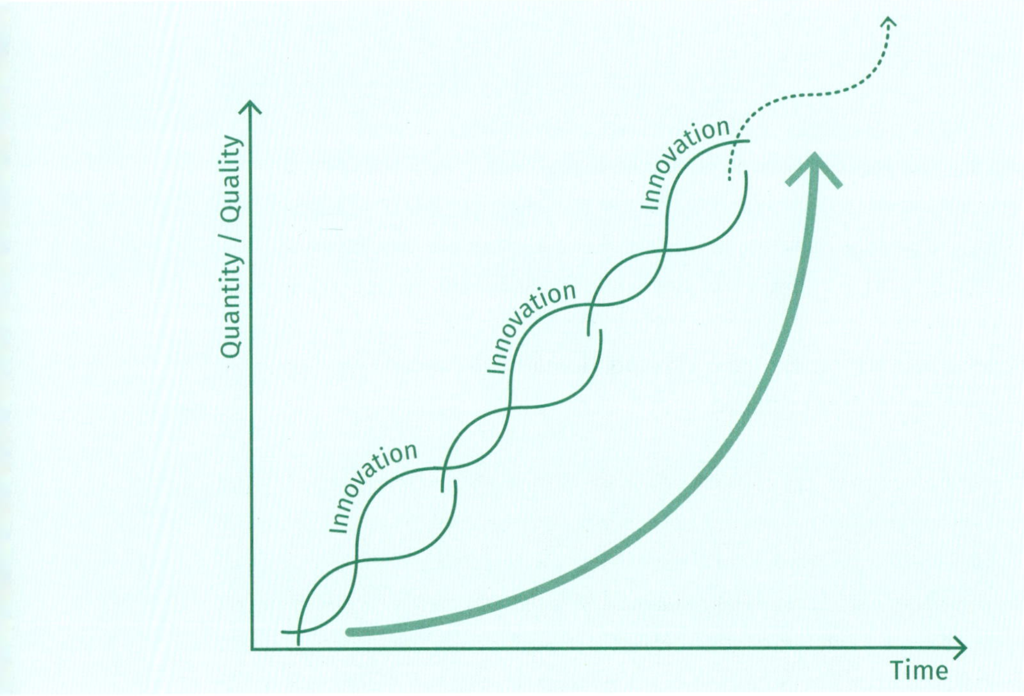

Figure 1: Innovation as a driving force in the market economy (based on Faix 2008: 20)

In simple terms: Entrepreneurs are forced to generate the new and the adapted in order to stand their ground in the face of competition, or even leave their competitors behind. As entrepreneurs do not (or cannot) permit a state of economic equilibrium to exist versus prospering companies, but rather strive to gain competitive advantage and then exploit this advantage, what this means for companies that "stand still" is that they automatically fall behind and in some circumstances are even ousted from the market.[1]

On this note, it is appropriate to consider recent developments in widespread globalization. The "break-off [...] away from the categorial framework of the national state" (Beck 1997: 13) further raises the scale and rate of this dynamic development. Companies now no longer only compete with neighbors in their own geographical region but also with companies from all around the world. The highly developed industrial nations in particular are continuously finding it more of a challenge to maintain their competitive standing in the face of nations with location advantages (intensity of regulation, wage levels, skilled workers, etc.). Even within specific sectors of industry, even on a microeconomic scale, globalization is resulting in more intense competitive pressure. In the past, long product life cycles influenced and shaped business strategy, and these long product life cycles made it possible for imitators to gain a foothold in markets where all the preparatory work had already been done, thus giving them the advantage of following in the footsteps of the innovator.

1 It has been observed time and again in an economy how new technologies spawn completely new sectors of industry (biotechnology, nanotechnology), fundamentally change others (smartphones, social media), or even eradicate them (e.g., in Germany: atomic power stations versus sustainable sources of energy).

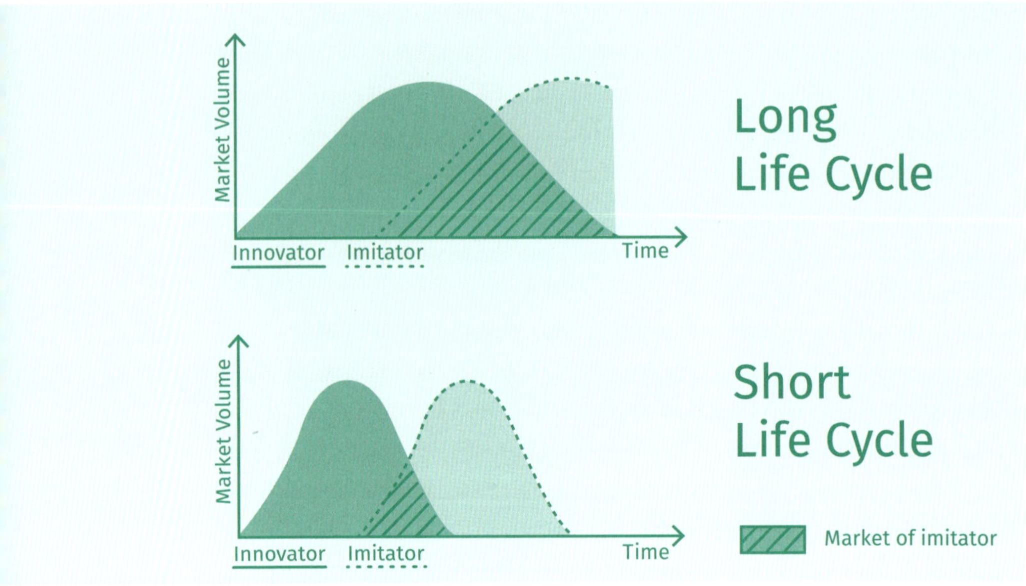

Figure 2: Market of the imitator (Nagel 1995)

Today, an imitator entering markets that have already been shaped by others has far fewer opportunities because product life cycles are distinctly shorter, noticeably affecting copycat behavior. It is only worth adopting the role of an imitator if the copycat product can be produced at a significantly lower cost. With a whole variety of products, competitive advantage is now measured in months. Accordingly, noticeably fewer small firms are now swallowed up by big companies. Instead, one increasingly observes slow, traditional companies being shoved aside by the fast and innovative companies. (Faix et al. 1994)

Another reason for the increased scale and rate of change is the rapid transformation into a (digitized) knowledge-based society. This is clearly reflected by the waning significance of classic production factors since the 1990s – land, work and capital – and as a consequence of this, substitution by another production factor: knowledge.

> *Money generates returns, but it does not think; machines can carry out undesirable tasks for people, but they do not invent things. As valuable as natural resources may be, and as large the army of workers may be, in the long term, they will not be able to compensate for knowledge generated by the individual. (Translation, Oelsnitz et al 2007: 37)*

Virtual mail-order businesses based on digital knowledge no longer have their own stocks. Customer orders are forwarded (practically in real time) to the manufacturers who ship the products directly to the customer. Progress based on digital knowledge in the field of bioengineering has made it possible to cultivate strains that are largely immune to pests and the weather. In the past, to generate similar returns, more land and more working hours had to be invested. Today, the relentless networking of sensors and actuators is resulting in knowledge that can not only be re-used to exploit product opportunities and raise productivity, but can also be used to "automatically" generate knowledge itself. The words of Friedrich List are gaining more and more in terms of validity: "The power of producing wealth is [...] infinitely more important than wealth itself." (List 1930 / 1841: 173) Given this new epoch of the "knowledge-based society," as it becomes easier to access this knowledge (in its digital forms, which means everywhere, at any time and in real time), and as people become better educated, it would be safe to assume that groundbreaking insights and inventions – and thus to a greater or lesser extent, major changes – will, at the very least, become more probable.

Ultimately, the scale and rate of change are additionally reinforced by three logically and causally independent processes, which in association with each other become mutually dependent. These processes are:

> technical or technological acceleration

> the acceleration in the pace of life

> the acceleration of social change

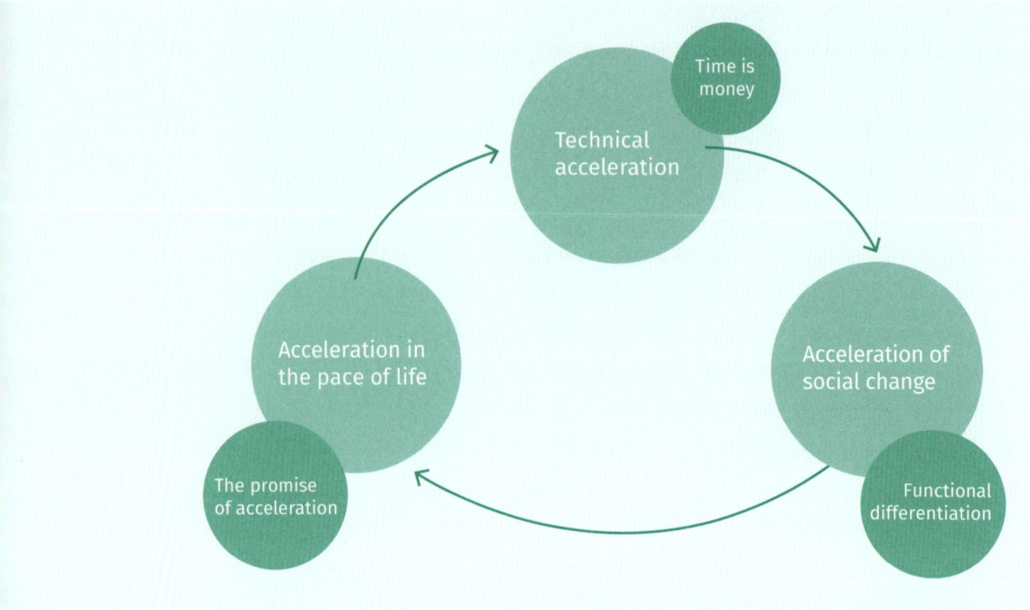

Figure 3: Reciprocal reinforcement in the triangle of change (Rosa 2008)

Technical acceleration refers in particular to areas such as transportation, communication and production. These days, distances are of little significance any more; with the Internet and with the physical support of high-speed modes of transportation, distances can be covered in a fraction of a second or several hours.

Acceleration in the pace of life is used to express the shortage of time – despite continual time savings through technology – resulting from the growth in tasks and possibilities. This growth is exponential compared to exclusively linear temporal acceleration.

Acceleration of social change is reflected in the accelerated forfeiture of action-based experience and expectations: As the rate of progress continues to accelerate, the half life of our acquired knowledge and insights – i.e., our knowledge of the world – is becoming shorter.

What this means for companies is that they can and must continually find answers or solutions to questions or problems that are new or have not yet been posed. This will have to be at shorter and shorter intervals. It is a never-ending task and many will be doomed to fail.

For the IBM CEO studies, a survey is conducted on current challenges by questioning entrepreneurs, CEOs and managers working throughout the world in the private economy and public sector. In the 2008 study, eight of ten respondents said they stood before significant change, although at the same time there was a major discrepancy in the ability to cope with this change. (IBM 2008: 7)

> *Constant change is certainly not new. But companies are struggling with its accelerating pace. Everything around them seems to be changing faster than they can. [...]Suddenly everything is important. And change can come from anywhere. CEO s find themselves — as one CEO from Canada put it — in a "white-water world." (IBM 2008: 15)*

Ultimately, the rafting alluded to can only be survived unscathed by not allowing oneself to drift aimlessly, but instead by actively making use of the inexorable change in potential kinetic energy by grabbing the rudder or paddles and being the first to guide the undamaged boat across the finish line. This can be achieved if managers run their business and activities by continuously "reading" or "sensing" the changing whitewater, and (once again) discover/ reinvent and optimize themselves. A company must react to continual changes in the world outside (which are sometimes only gradual, but sometimes also radical) by changing the world within - gradually, but sometimes also radically - as well as changing course. Things that are new or unlike anything ever encountered in one's surroundings, have to be countered by something revolutionarily different and evolutionarily superior.

Given the possibility of change, he who preempts change in himself is primarily the one who gains the advantage by proactively bringing about something new in the inside world to react to something new in the outside world. Given the reality of change, only he who changes has a future, he who reacts to something new in the outside world with something new in the inside world. The willingness, ability and readiness to counter possible or actual change creatively have always been a requirement and a yardstick of the future-readiness of organizations and enterprises, of states and economies, but also of individuals. In a nutshell, what this means is that "in any epoch of rapid change, those organizations [as well as states and economies, but also individuals, that are] unable to adapt are soon in trouble, and adaptation is achieved only by learning" (Revans 1983: 11). Or as Charles Darwin wrote in On the Origin of Species:

[The ability to adapt better to changing habitats may determine] which individuals shall live and which shall die,— which variety or species shall increase in number, and which shall decrease, or finally become extinct. [...] The slightest advantage in certain individuals, at any age or during any season, over those with which they come into competition, or better adaptation in however slight a degree to the surrounding physical conditions, will, in the long run, turn the balance. (Darwin 507-508)

In simple terms: In a world which is increasingly shaped by the fact that everything is constantly on the go – and is sometimes even on the verge of chaos – entrepreneurs and companies have to be prepared, at any point, to dare to make a development "spring forward" themselves – as an evolution or even as a revolution. The "genes" this takes must be within the entrepreneur; companies must be led in this way by the managers.

Today's CEOs face grueling conditions. Buffeted by volatility, they have come to expect the unpredictable. But they know that the return to growth will require more than resilience or sure footing. They need to spring forward with the vigor of Olympic-caliber athletes. (IBM 2010: 52)

A daring undertaking (i.e., action) should in itself be tantamount to the guiding principle of an entrepreneur, but also of managers who act in an entrepreneurial manner. Conversely, a crisis or a loss of competitiveness are "the results of a novelty deficiency. [...] For the innovative, salvation lies in the future of the not-yet-witnessed." (Gronemeyer 2000: 6) Much more than in the past, companies will need a self-conception of creativity in the future, and in particular a willingness and an ability to create something new and different – or at least allow this to happen.

To survive in a world of perpetual renewal, the categorical imperative, the primary entrepreneurial principle, lies in the ability and the willingness to innovate.

Continually not just thinking of the New, but also allowing it to become a value-creating and thus also value-adding reality – this is what dictates the fate not just of an economy, but also of companies (who support it).

The importance of creative destruction 1.2

Irrespective of a company's size, on a fundamental level there are two possible options for a firm to safeguard or build on its competitiveness, and these can be combined.

> The first option spans a variety of activities relating to mergers and acquisitions (M&A), so these could be spin-offs, spin-outs, collaboration (such as joint ventures, strategic alliances, virtual enterprises, research networks), stakeholdings, takeovers and mergers (such as spin-ins, range extensions/completion), restructuring and hedging (such as going public, buy-outs), and sell-offs and liquidations (such as streamlining or spin-off liquidations).

> The second option is to come up with new ideas oneself, and to be as innovative as possible in doing so.

Which of these two options is most likely to result in success? According to the results of an analysis carried out on Fortune Global 500 companies (Raisch, Probst, Gomez 2007), innovations make a significantly higher contribution to added value than acquisitions: "Most companies enjoying sustainable growth have an organic growth strategy, in which acquisitions [or general M&A activities] play a secondary role. The primary aim is to grow more quickly than the competition through one's own resources." (Ibid. 44)

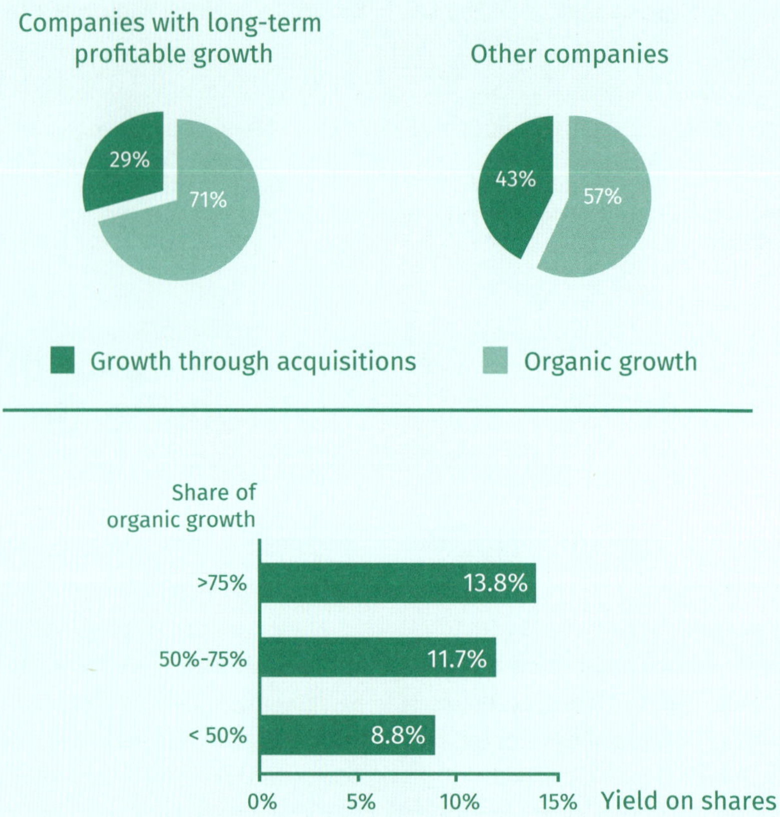

Figure 4: Relationship between organic growth and growth through acquisitions at Fortune Global. 500 Companies (1995–2004) (Raisch, Probst, Gomez 2007: 43)

M&A activities are not wrong: Profitable companies place a clear emphasis on innovation in the long term; at the same time, they use M&As as a supplementary activity in two ways:

1. Successful companies enter new markets (e.g., through acquisitions) to quickly attain competitive mass and occupy a position in the market.

2. Successful companies buy companies or acquire a stake in companies with a strong track record in innovation. (Ibid. 43 ff.)

On examination, when it comes to small and medium-sized enterprises, we subscribe to the following view: "Growth through acquisitions is the second-best strategy. The best is organic growth. We place emphasis on innovation and see acquisitions purely as a complement." (Translation of Henning Kagermann, Chairman of Executive Board at SAP AG, quoted in ibid.: 40, representing a major company, but one that started as a microbusiness and attempts to remain innovative despite its size).

A number of theoretical growth models have been developed in recent years with the aim of explaining growth as the result of innovation.[2]

Such models have been empirically substantiated by a series of studies.

- Smolny and Schneeweis (1999: 468) came to the conclusion that: "Innovators exhibit significantly more favorable developments in turnover and employment than non-innovators; on a corporate level, both product and process innovations led to higher turnover and higher employment."[3]
- Innovations also appear to play a signification role in export:

  Using the innovation impulses and obstacles as instruments for actual innovation, we find that innovation emanating from the variation in these impulses and obstacles leads to a share of exports in firms' total turnover that is roughly 7 percentage points higher on average. Given a mean export share in our sample of roughly one quarter, this is a substantial effect. Therefore, our results support the prediction of the product-cycle models that innovation is a driving force for industrialized countries' exports. The effect is heterogeneous across sectors, hardly detectable in relatively traditional sectors and as large as 17 percentage points in the relatively modern sectors of the German manufacturing economy. [...] Being innovative causes firms to have substantially larger export shares than non-innovative firms in the same sector. (Lachenmaier, Woessmann 2004: 24 ff.)

2 Cf. e.g. Aghion, Howitt (1998) or Barro, Sala-i-Martin (2004).
3 It is not possible at this point to say whether this effect is directly dictated by innovations or other factors. Another explanation could be, for example, the so-called Matthew Effect (Merton 1968), which can be summarized by a quotation from the bible: "Unto every one that hath, shall be given." It is possible that the success of some companies (e.g., when fighting for the most talented people) stems not from the fact that they are currently particularly innovative, but rather that they earned themselves the reputation of a particularly innovative company in the past.

> For the employment market, innovation means this:

So, to sum up, almost all of our innovation measures show a significantly positive [and robust] effect on employment. Surprisingly, this effect is higher for process innovation than for product innovations. (Lachenmaier, Rottmann 2007a: 21)

Our analysis gives strong evidence that innovations have a significantly positive effect on employment growth in German manufacturing firms. This is true for both types of innovations: for the introduction of product innovations as well as for the implementation of process innovations. Process innovations showed a higher effect on the employment growth rate than product innovations in most cases. (Lachenmaier, Rottmann 2007b: 20)

> This economic and entrepreneurial significance of innovation appears to be irrefutable and evident:

[...] our results are supportive of the suggestion of the product-cycle trade models that industrialized countries may have to continually innovate if they want to remain competitive on global markets and maintain their living standards. (Lachenmaier, Woessmann 2004: 25)

Innovation is widely considered as the life blood of corporate survival and growth (Zahra, Covin 1994: 183)

Innovation represents the core renewal process in any organization. Unless it changes what it offers the world and the way in which it creates and delivers those offerings it risks its survival and growth prospects (Bessant et al. 2005: 1366)

If [...] companies fail to innovate, jobs and profits will suffer, and our standard of living will fall compared with other countries" (Department of Trade of the UK 2003)

This connection between innovation and affluence is of immense importance for all kinds of economies; for industrial nations, the connection is existential.

Although substantial gains can be obtained by improving institutions, building infrastructure, reducing macroeconomic instability, or improving human capital, all these factors eventually seem to run into diminishing returns. The same is true for the efficiency of the labor, financial, and goods markets. In the long run, standards of living can be enhanced only by technological innovation. Innovation is particularly important for economies as they approach the frontiers of knowledge and the possibility of integrating and adapting exogenous technologies tends to disappear. Although less-advanced countries can still improve their productivity by adopting existing technologies or making incremental improvements in other areas, for those that have reached the innovation stage of development this is no longer sufficient for increasing productivity. Firms in these countries must design and develop cutting-edge products and processes to maintain a competitive edge. (WEF 2011: 8)

For emerging economies striving to make the leap and become an industrial nation, but also for industrial nations themselves, it is valid to say that positive economic developments are predominantly achieved through innovation. Conversely, a standstill or even a slackening in the ability to innovate and carry out innovative activity has a negative impact on economic development.

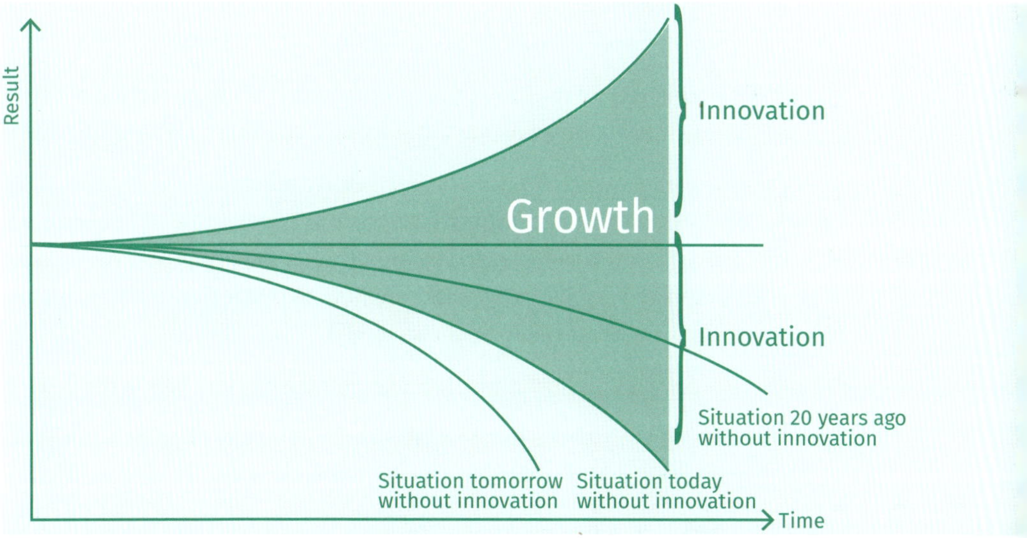

Figure 5: Growth through innovation (Faix 2008: 20)

These results are reflected in the decisions made by managers at the companies who were surveyed as part of the IBM Global CEO Study 2008: Two thirds of respondents adapt the business model of their company and innovate on a major scale. (IBM 2008a: 7) The authors of the study conclude from the statements that this, among other factors, characterizes the company of the future: (cf. in the following, IBM 2008a: 8 f. and 54 f.)

- The company of the future is in a position to change quickly and successfully.
- The company of the future exceeds the expectations of its increasingly demanding customers.
- The company of the future radically questions its previous business model – even if it is still profitable – and uses this to lay new foundations for competition.
- The company of the future changes continuously, but because of its company culture, employees have no problems with the resulting unpredictability.
- The company of the future is the right place for visionaries – people who question inherited assumptions and suggest radical alternatives, even if implementing these alternatives may appear impossible at first glance.
- The company of the future puts processes and structures in place that promote innovation and transformation in the company.
- The company of the future thinks creatively and unconventionally. It inspires innovation by imagining starting right from the beginning again.
- The company of the future inspects other industries exactly because it knows that groundbreaking ideas spread like wildfire. It keeps an eye open for customer and technology trends that change other market sectors and segments, considering how these trends could be applied to its own industry and business model.
- The company of the future often tries out business models in the market.
- The company of the future manages the company of today while it experiments with the business model of tomorrow.

Issues, aims and approach 2

The value and benefit of innovation for companies and economies is fundamental – and it is difficult to stress enough how fundamental. Surprisingly, however, there is no quantitative variable or measurement to express the value or the significance of an innovation.

Naturally, there have already been scores of attempts to capture the phenomenon of innovation in a more differentiated manner. First and foremost, one thinks in this context of the distinction made between radical and incremental innovation. In essence, these concepts primarily capture the degree of novelty or the extent of dissimilarity which make a given innovation different from existing ones. However, such concepts for differentiating innovation basically say nothing about the value or significance of "the New," or, in a word, about the quality of an innovation. As a result, what is required in this discussion is a new variable, one which reliably reflects the quality of an innovation. We call this quantitative variable InnovationQuality.

This raises the following issues:

> What is, or what constitutes InnovationQuality?
> How can the quality of an innovation be expressed or even perhaps measured?
> How can one influence InnovationQuality in a targeted manner and plan this; that is, what would an InnovationQuality Management model look like?

In simple terms, our aim is to successfully bring together the concepts of innovation and quality, to merge two terms with one another in a way in which they have not been brought together before. Interestingly, if this succeeds in becoming a reality and it creates value, this idea in itself is an innovation. This is because, in its essence, that is what innovation is about: "To produce other things, or the same things by a different method, means to combine these materials and forces differently. [...] Development in our sense is then defined by the carrying out of new combinations." (Schumpeter 1934: 65-66) According to Schumpeter, as a first step, an innovation thus results from a combination of the existing in a new or different way. What is important is any particular success that results from this, which we are optimistic enough to assume will be the case.

But how exactly does this combination work, or what does one actually do when one innovates? To examine this, we go on a short thought excursion into the theory of metaphors.

A metaphor is an expression in which a term or a number of terms are combined with another term or number of terms. According to (structural) semantics, a term implies a number of distinctive characteristics.

Figure 6: The term "Lion" as a group of distinctive characteristics

A metaphor (i.e., a combination of terms) is given when a characteristic of Term A is transferred to Term B.

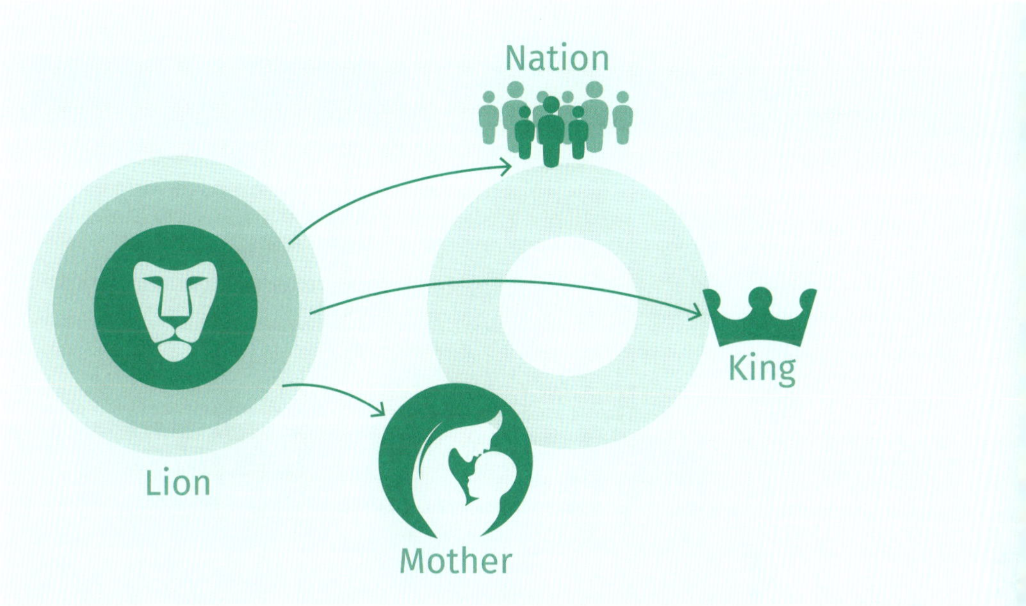

Figure 7: A metaphor and its role in transferring characteristics

From this, the following metaphors can be derived.

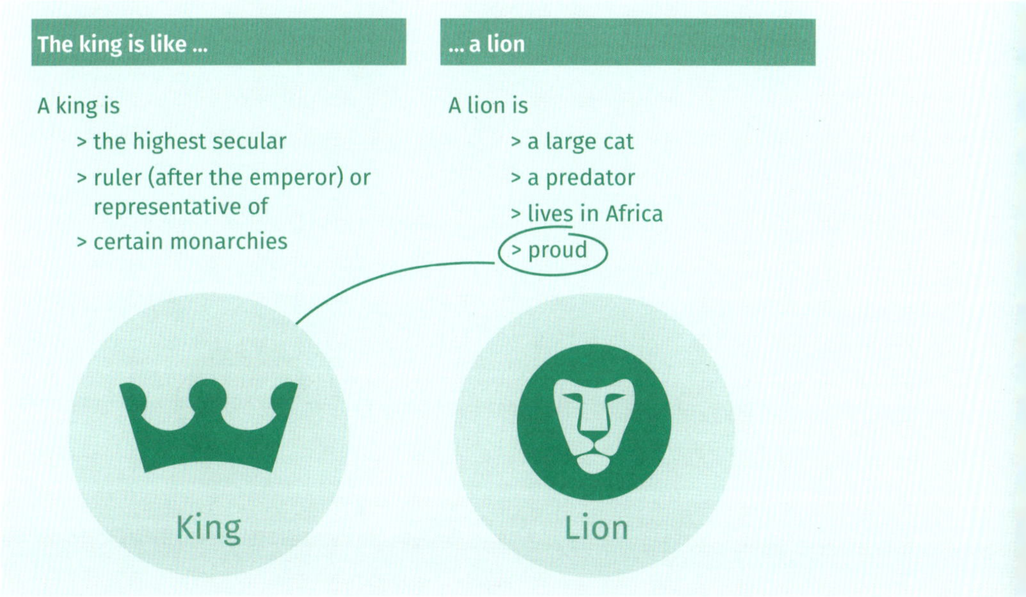

Figure 8: Metaphor, "if one compares a king with a lion, then ..."

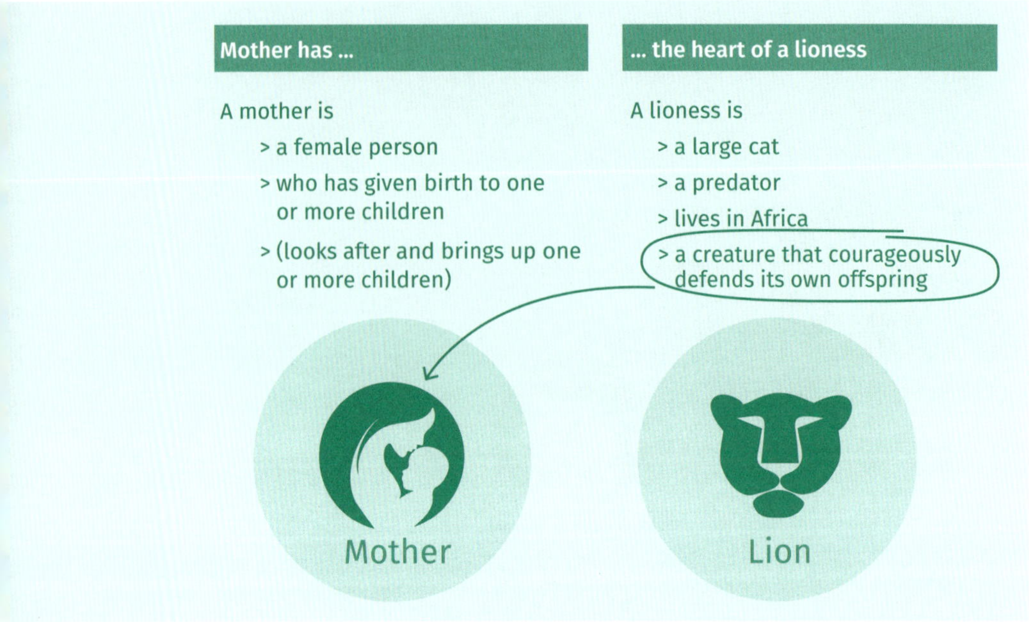

Figure 9: The metaphor, "if one compares a mother with a lioness, then..."

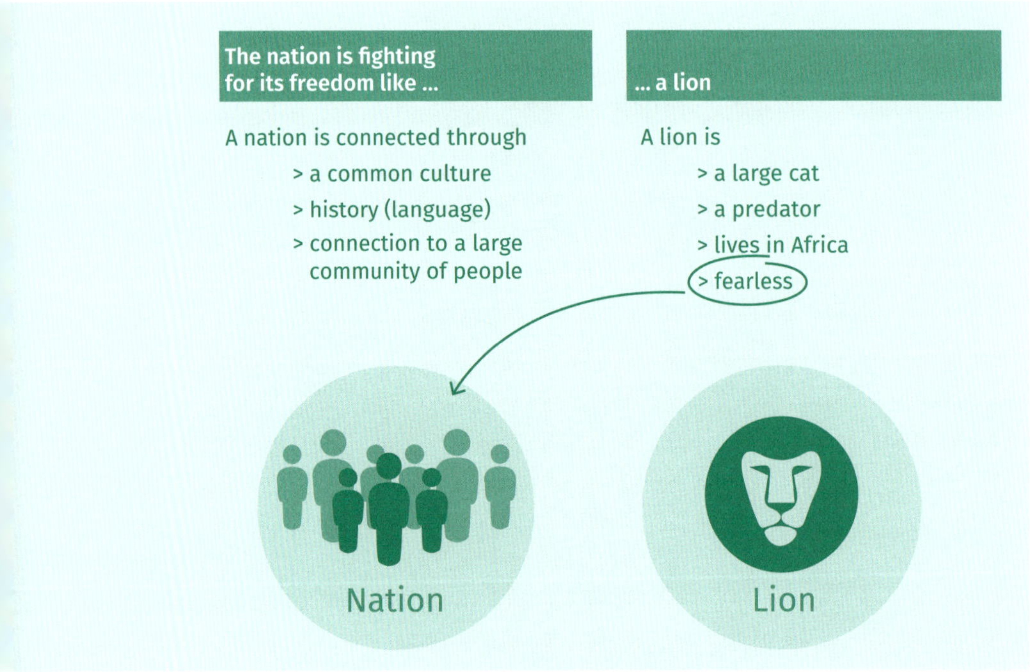

Figure 10: The metaphor, "if one compares a nation with a lion, then..."

To use or understand a metaphor, one first requires an interpretation, namely:

> An interpretation of the distinctive characteristics of the original concept (e.g., king, mother)
> An interpretation of the distinctive characteristics of the associated concept (e.g., lion).

It is only when this understanding exists that characteristics can be transferred in such a way that terms can be connected. Applying this idea to our term, InnovationQuality, this means: For these two terms to be combined, an understanding is needed of their distinctive characteristics.

One would assume that it should not be a problem to understand the term "quality" since there are standardized ISO benchmarks which lay down the distinctive characteristics of "quality." But it will be more difficult with the understanding of "innovation." Of course there are a (not insubstantial) number of authorities, some widely recognized, who have defined "innovation." But in contrast to "quality," there are no standards, no benchmarks to make a more detailed differentiation of the term "innovation." Exaggerating in order to make a point: "Quality" appears to be more of a domesticated phenomenon, whereas the phenomenon of "innovation" still appears to be somewhat undomesticated.

Our approach is broken down into five steps:

1. Introduction to the two terms of "Quality" and "Innovation" by closely examining the phenomena and their distinctive characteristics
2. Combination of the two terms, or the two phenomena of "innovation" and "quality"
3. Specific definition of the interpretation of InnovationQuality, also as a quantitative variable
4. Outline of a model for InnovationQuality
5. Illustration of an approach for managing InnovationQuality

For the first step, we illustrate the two terms, or phenomena, and their distinctive characteristics, and, in doing so, examine the understanding required to create this combination.

As the second step, we merge the two terms or rather, we transfer distinctive characteristics of one phenomenon to the other. The result of this transferal or merging process should, by all accounts, establish an extensive, general and pragmatic interpretation of the term "InnovationQuality." The word *extensive* is used here to mean that innovation is a multifaceted phenomenon with diverse meanings, and that all these colorful characteristics are also contained within the term InnovationQuality. We use *general* in part to indicate that, although we examine the phenomenon of "innovation" from the perspective of companies in this book, our interpretation of InnovationQuality should be compatible with non-business contexts and thus explicitly also include social innovation. By *pragmatic*, we mean that InnovationQuality should fundamentally be an understandable and realizable concept; "pragmatic" should not be interpreted to mean that we are using the term "InnovationQuality" to provide a readymade recipe that can be applied immediately to a particular situation.

Our third step is to make our interpretation of "InnovationQuality" more concrete, to even encompass a quantitative variable. With this step, too, the outlined formulae should fundamentally provide extensive, general and pragmatic access and be compatible in non-business contexts.

For the fourth step, we use this interpretation of InnovationQuality to outline a model which – drawing on the example and referring to the existing model of the European Foundation for Quality Management (the EFQM Model) – can be used to manage the quality of innovation.

Our fifth step is to illustrate an approach for managing InnovationQuality.

In conclusion, to lay the foundations for further discussion, which we would welcome, we provide suggestions and ideas for dealing with the processes involved in the antithetical phenomenon of InnovationQuality.

To conclude this introduction, here are some "usage instructions": The results of our work, as described in this book, and our discussions are outlined in order to explain how we arrived at the term InnovationQuality. These should serve to provide an initial catalyst for targeted entrepreneurial transfer, scientific discussion, differentiation, developments and the like. We thus make no assertions in terms of definitiveness and conclusiveness.

Step 1:

Introduction to the two terms of "Quality" and "Innovation" by closely examining the phenomena and their distinctive characteristics

Step 2: Combination of the two terms, or the two phenomena of "innovation" and "quality"

Step 3: Specific definition of the interpretation of "InnovationQuality", also as a quantitative variable

Step 4: Outline of a model for InnovationQuality

Step 5: Illustration of an approach for managing InnovationQuality

A closer examination of the phenomenon of "quality" 1

Although, as indicated in the introduction, the phenomenon of quality is somewhat more "domesticated" than the phenomenon of innovation, this by no means implies that there is wide-scale unanimity regarding the actual nature of "quality." But unlike innovation, there is at least a generally recognized definition for the phenomenon of quality: A standard definition has been in place since December 2005 under the quality management standard ISO 9000:2005-12. According to this standard, quality is defined as "the degree to which a set of inherent characteristics fulfills requirements."

As widely recognized as this definition may be, there are a variety of ways to interpret its predicators (key sentence elements). So it is already possible to conclude that the phenomenon of quality is only seemingly "domesticated," in other words: Although there are a myriad of standards and definitions to describe quality, some rather heterogeneous and in part vague interpretations of "quality" exist.

The "degree" of quality 1.1

The predicator "degree" describes a gradation of the existence of an attribute, of a condition or state. Essentially, in quality management a distinction between three dimensions of "degree" can be made:

1. Degree meaning different levels of the existence of a structure or of potential; that is, the nature of the quality of frameworks consisting of elements (products, services, etc.) and the nature of the capability to create these frameworks.

2. Degree meaning different levels of the existence of a process; or rather, the nature of the quality of procedures used to create structures or results.

3. Degree meaning different levels of the existence of a result; in other words, the nature of the quality of effects, end status, yields, etc., resulting from the existence of structures and processes.

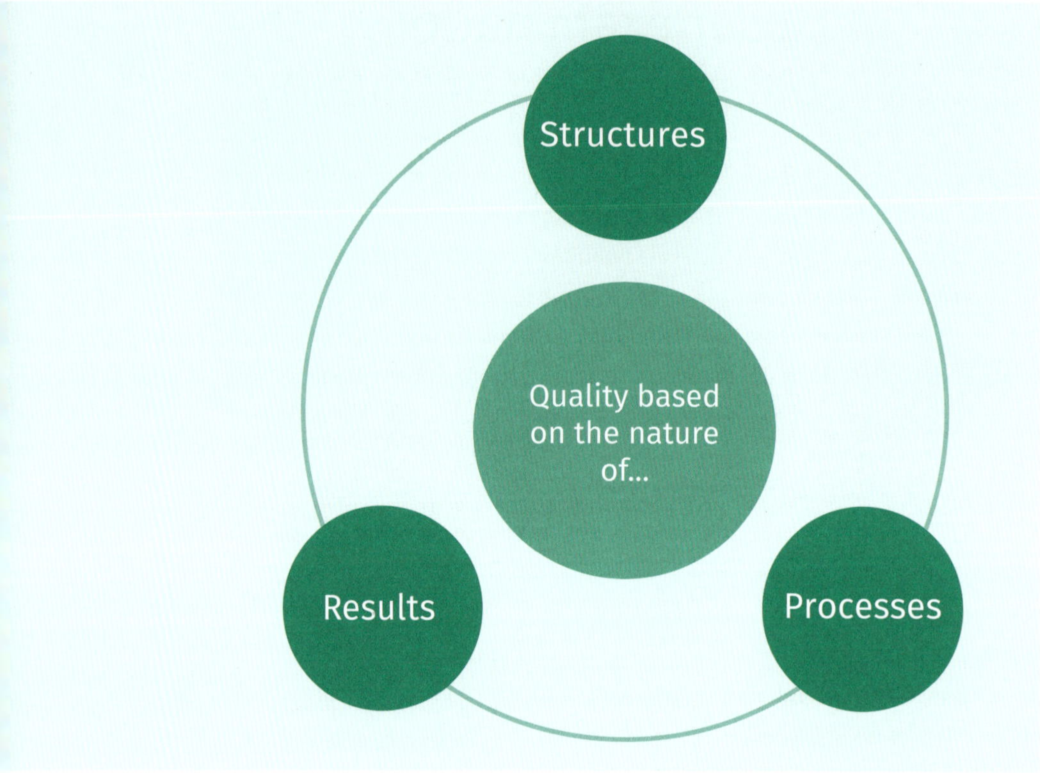

Figure 11: Dimensions of the nature of quality

1.2 The "set of inherent characteristics" of quality

The predicator "set of inherent characteristics" describes the distinct features of specific objects. The nature of an object can be outlined in quality management through a three-dimensional, multi-factor third-order construct. Multi-factor means that this construct cannot be expressed directly through manifest variables; instead, this construct is derived from a number of so-called latent variables. This makes it necessary to define a concept – in an iterative, step-by-step process – to describe the nature of objects on the highest level (construct level) and on the mid-level (dimension level), right down to the lowest level of abstraction (factor level). (cf. Bruhn 2013: 44)

Such a process for defining the concept of quality could work like this:

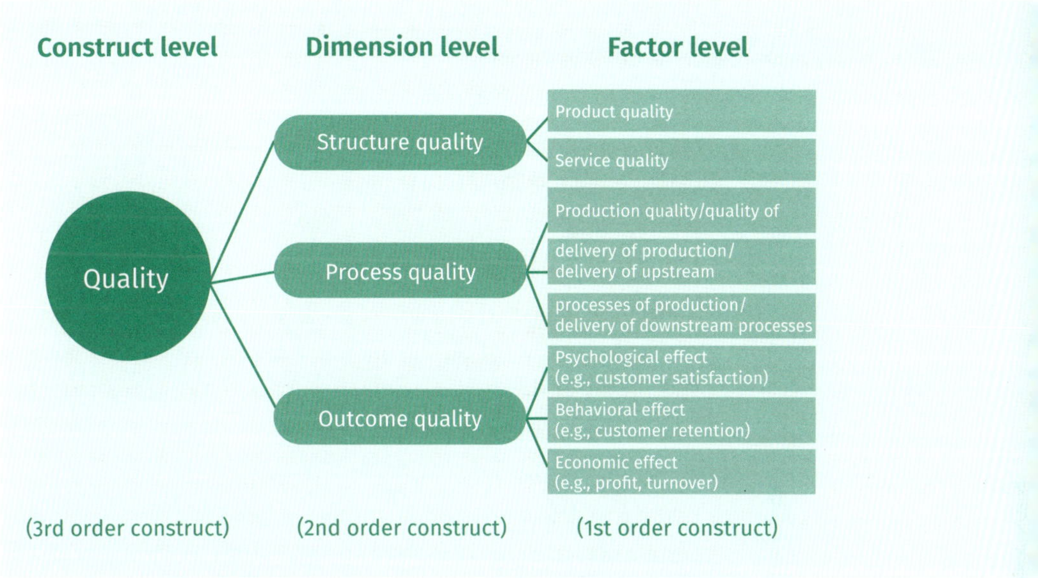

Figure 12: Quality as a three-dimensional, multi-factor third-order construct

Accordingly, distinctions in the construct of "quality" can be made with the following dimensions and factors, or, in other words: It can be shown if something "is high-quality" or is "of a certain quality" through the following:

> Structure quality – the nature of a framework:
  - Product quality: the nature of a produced item or object
  - Service quality: the nature of a service that is not directly intended for the production of objects

> Process quality: the nature of a procedure that is indirectly or directly related to the production of an object or the delivery of a service
  - Production quality/delivery quality: the nature of procedures which are directly related to the production of an object or the delivery of a service
  - The production/delivery of upstream processes: the nature of imperative or non-imperative procedures, which take place before the production of an object or the delivery of a service

    - The production/delivery of downstream processes: the nature of imperative and non-imperative procedures, which take place after the production of an object or the delivery of a service

> Outcome quality: the nature of an impact resulting indirectly or directly from the production of an object or the delivery of a service

    - Psychological impact: the nature of the opinions, attitudes, convictions and beliefs of a recipient toward certain objects or services from an originator, specifically or toward the originator in general (e.g., customer satisfaction, image)
    - Behavioral impact: the nature of the witting or unwitting reactions of a recipient toward certain objects or services from an originator, specifically or toward the originator in general (e.g., customer loyalty, purchase propensity)
    - Economic impact: the nature of entrepreneurially relevant factors relating back to witting or unwitting reactions of a recipient toward certain objects or services from an originator, specifically or toward the originator in general (e.g., profit, turnover)

1.3 The "requirements" relating to quality

The predicator "requirements" describes the expectations held by a person (the "subject") regarding specific objects. "Requirements" thus lay a foundation, which the subject draws upon to evaluate the quality of specific objects.

Within this context, a distinction must first be made to ascertain who the person (subject) actually is, who has the requirements and makes an evaluation. At a fundamental level, a distinction can be made here between the manufacturer of an object ("producer") and the recipient of this object ("customer"). The requirements of these two parties can be, but do not have to be, consonant (or match): Perhaps a customer has different requirements regarding faultless, wear-free operation or the longevity of a technological object compared to the producer.

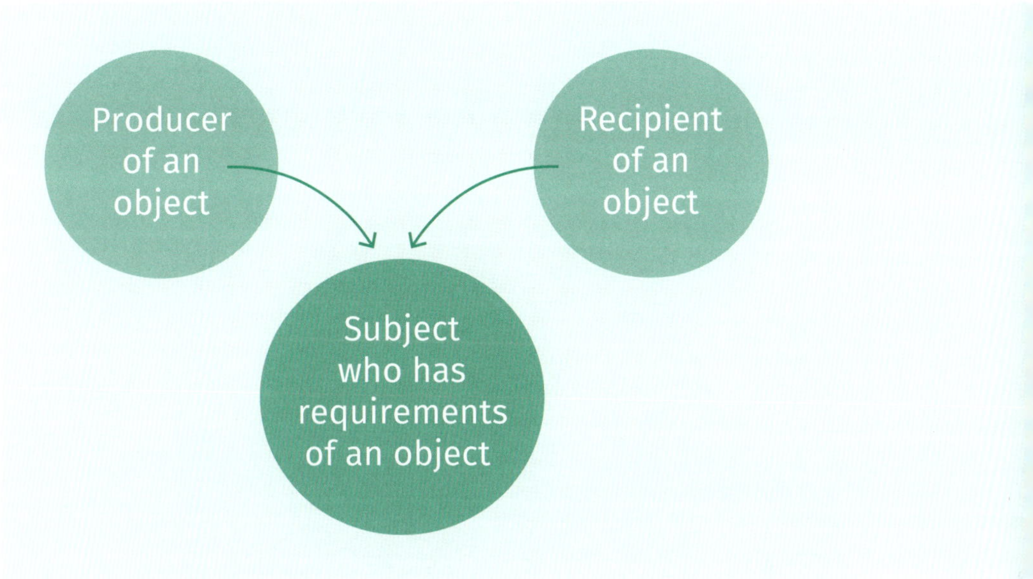

Figure 13: Determination of a person ("subject") who evaluates the quality of an object

Furthermore, with requirements, a distinction must be made between objective and subjective requirement criteria. The starting point for objective requirement criteria is a set of explicitly nameable and identifiable characteristics. These are frequently captured by quantitative variables and indicator systems ("quality manual"). The starting point for subjective requirement criteria is a set of psychological conditions or states in the form of expectations. In this context, distinctions can be made between:

- Basic expectations ("must criteria"): what the recipient (e.g., "customer") requires of an object on a fundamental level
- Delivery expectations ("should criteria"): what the recipient considers probable, given comparisons with similar objects
- Enthusiasm expectations ("can criteria"): what will surprise the recipient in a positive sense, given comparisons with similar objects

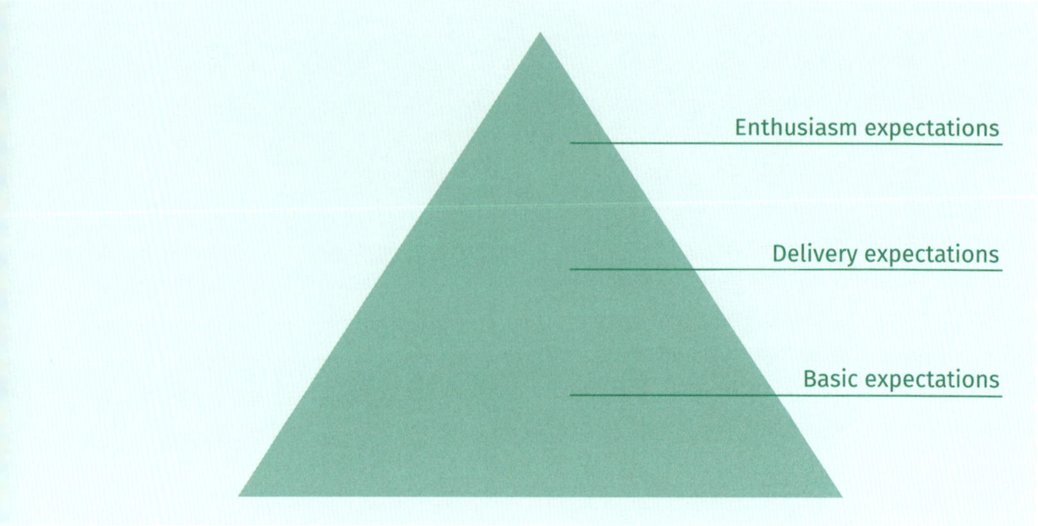

Figure 14: Subjective requirement criteria ("expectation") relating to the quality of an object

Distinctions can also be made between requirements depending on whether an object is judged specifically or holistically. When an object is judged specifically, individual indicators or parameters are used (e.g., how long a rechargeable battery lasts or the speed of a processor in a smartphone). When an object is judged holistically, all components relating to an object are merged into a single entity and are evaluated as such (e.g., this smartphone is great or it is just what I need).

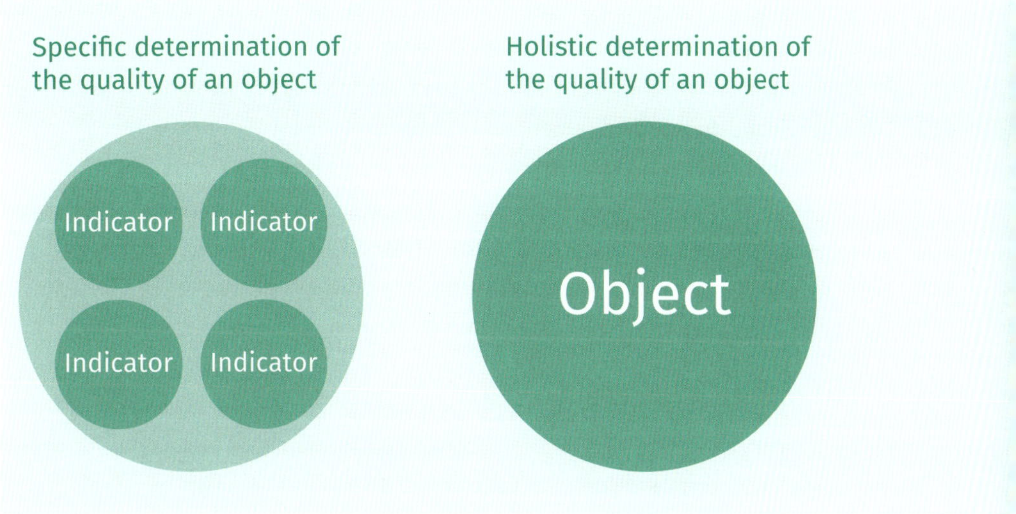

Figure 15: Specific and holistic determination of the quality of an object

Finally, a distinction can be made between requirements by examining whether someone draws on absolute or relative criteria relating to the quality of an object. With the absolute determination of quality, an object is categorized according to different classes such as "good, average, poor." With relative determination, an object is either judged according to aspects such as value for money (price-performance ratio) or a judgment is based on comparable objects.

2 A closer examination of the phenomenon of "innovation"

Even if the social sciences and, in this context, the economic sciences have been discussing "innovation" at least since the times of Schumpeter, and this has intensified over the last 20 years (see also authors such as Drucker 1981), innovation remains a phenomenon that is still worthy of detailed discussion. A particular feature of social scientists is that they have a sublime tendency to be at odds with one another, particularly when it comes to fundamentals. This also, or above all, stems from the fact that the disputed terms that such scientists work on are not words like "chair" or "table," which can be pointed at and measured; such words at least evoke phenomenologically comparable concepts in different people (chair = subject can sit on it; table = objects can be placed on it). As a result, if someone involved in a discussion of these concepts actually doubts the fundamental existence of the terms, one can literally shove an object in front of them or bang it on their head. The term "innovation" is thus a typical term of the social sciences. "Innovation" somehow points to a reality, to an existence that is conceivable and must thus be expressed in a conceivable way; however, the essence – the actual content of the term, its actual meaning – requires detailed discussion.

Also, and especially with the phenomenon of innovation, aside from this fundamental issue, there is the following epistemological condition: According to the interpretation of so-called operative constructivism, as developed by Niklas Luhmann, observations have no immediate access to an objectively existent world.

> *[One can thus] neither assume that there exists a world at hand consisting of things, substances, and ideas, nor can one designate their entirety (universitas rerum) with the concept of a "world." For sense-systems, the world is not a giant mechanism that produces states out of states and thus determines the systems themselves. The world is rather an immeasurable potential for surprises, it is virtual information that needs systems to produce information, or more precisely; to ascribe to selected information the sense of being information. (Translation, Luhmann 1997: 46)*

We find neither objects, nor terms, nor questions in the world. It is, and remains, an entirety out of which phenomena only emerge when people turn their eye to it.

> *Cognition is not like the environment, as the environment contains no distinctions, it simply is as it is. [...] Everything that can be observed is a personal action of the observer, [...]. So there is nothing in the environment that corresponds to cognition; since everything that corresponds to cognition depends on decisions, according to which something is distinguished as this and not that. In the environment there are thus neither things nor events if such a term should denote that something, that is described as such, is different from something else. (Translation, Luhmann 1988: 15 f.)*

Applying this to "innovation" as an economic phenomenon, this means: The world of business is, at first glance, an entirety that is actually inseparable. Within this entirety, we observe "things" and name, for example, one thing an innovation and not the other ¬– i.e., we make distinctions between different elements within the entirety of the "world of business." These distinctions do not exist "naturally," they are profoundly a personal cognitive process within the observer; another observer may fundamentally observe the phenomenon of "innovation" differently and consider it to be something fundamentally different – naturally, and specifically, this is also valid for all of our definitions and models, and this includes the term InnovationQuality!

This inextricably epistemological issue is summarized by Damanpour and Schneider (2006: 216) thus: "Innovation is studied in many disciplines and has been defined from different perspectives." Naturally, against the backdrop of a myriad of attempts made by scientists and business practitioners to define the term "innovation," there are differing degrees of overlap in terms of content. Naturally, there are a good number of authorities who prescribe a pragmatic approach. So one can point on the one hand to international organizations such as the OECD with its Oslo Manual ("Guidelines for collecting and interpreting innovation data") or the World Economic Forum (WEF), which, in its Global Competitiveness Report, among others benchmarks the ability of countries and economies to innovate. On the other hand, one can point to luminaries and "legends" of innovation research, especially the "inventor" of the term innovation, Joseph A. Schumpeter. And finally, one can point to innovation system reports such as the Innovation Union Scoreboard of the European Union, the annual report of the German Federal Government's expert research and innovation commission, plus a variety of other domestic reports.

Nevertheless, there is no generally applicable and irrefutable definition of "innovation" because even these innovation definitions produced by the aforementioned authorities can be questioned and criticized. As early as 1984, Ettlie et al. observed that this vacuum is a problem not just for research but also in business practice. The fact that this situation has still not changed to this day is reflected in a comment made by Adams et al. (2006: 22): "the term 'innovation' is notoriously ambiguous and lacks either a single definition or measure."

In our closer examination of the phenomenon of "innovation," our approach is in keeping with the philosophical school of thought of pragmatism. This primarily implies that all consideration should be based on the primacy of practice. Accordingly, the meaning or the validity of terms is illuminated by practical reference, or, expressed another way: The meaning attached to a term depends on how people use the term in everyday situations. The reality of a term stems from the fact that people observe a phenomenon in their Lebenswelt – their everyday life – and for this phenomenon, if they stop to think or try to exchange notes with other people, they use a certain terminological placeholder. To prevent this pragmatic approach from venturing into the unscientific, we examine the phenomenon of "innovation" more closely by asking: How do scientists and business practitioners use the term "innovation" and what do these people see in their mind's eye when they talk about the term? The result will, of course, not provide us with a definition of innovation but we do at least calibrate our own observation and gain pointers for the properties of this phenomenon named "innovation."

Baregheh, Rowley and Sambrook (2009) conducted a content analysis and arrived at around 60 definitions for the term "innovation." The aim of the study was to capture the properties that are the constituent elements of the term "innovation." Most of the definitions that were examined originated from publications dating back to the period between the 1930s and the latter part of the first decade of the current century, with an emphasis on organizations and, as a function of this, companies in particular. The authors openly admit that time and disciplinary limitations had been obstacles for a generally valid definition.[4] Conscious of this, but given the comprehensive nature of the literature survey, the findings of this study can, however, be considered a fruitful first step. In a nutshell, the approach taken by the authors was to investigate the definitions they had identified to see whether and how often certain terms were mentioned. The most frequently named terms were then clustered and finally synthesized into distinct characteristics, which appear to be the constituent elements of an innovation. In the following, our aim is to decipher the three essential, distinct characteristics of the phenomenon of "innovation" and to summarize this:

4 It is also worth noting that the authors primarily referred to publications from authors in the English-speaking world; one possible assumption is that authors from other cultural or linguistic circles use different definitions.

> The nature of innovation: A decisive feature of an innovation lies in the fact that, in one way or another, it represents something new or something that has changed.

> The social context of innovation: An innovation is the realization of an idea in a specific object which has been introduced within a social context (e.g., a "market"). The introduction has implications within this social context.

> The type of innovation: Something new or something that has changed can be reflected in, or expressed through, different services or outcomes.

2.1 The nature of innovation

Joseph A. Schumpeter described the nature of an innovation as: "The doing of new things or the doing of things that are already done, in a new way" (Schumpeter 1947: 151). An innovation therefore has in its innermost essence something to do with "the New." But what actually is "the New?" And who actually decides or perceives whether something is new or not? On a fundamental level, a distinction can be made between two interpretations of this:

1. The originator of an object decides/perceives whether it is new: "The minimum requirement for an innovation is that the product, process, marketing method or organization method must be new [...] to the firm." (OECD 2005: 45)[5]

2. The recipient of an object decides/perceives whether it is new: "An innovation is an idea, practice, or object that is perceived as new by an individual or other unit of adoption." (Rogers 2003: 12)

For example, with respect to a product innovation, for the producer "new" would mean that a fundamental change is made to something that is being produced, such that the nature of the product changes, such that there is a fundamental change at the nucleus. For the producers of cars, naturally a com-

5 Cf. also: "An innovation can be new to the world, or new to a sector or market, or new to an agent. [...] Innovation also occurs when a firm introduces a product or process to a country for the first time. It occurs when other firms imitate this pioneering firm. Moreover, it occurs when the initial or follower firms make minor improvements and adaptations to improve a product or production process, leading to productivity improvements." (Dutta 2011a: 4)

bustion engine or a fuel cell is something new. Naturally, for the recipients, the switch made with cars from gasoline or diesel engines to alternative drive systems appears to be just as significant and radical with respect to their everyday lives. However, for the recipients, such developments are always just a different answer to one and the same question: How can I get from point A to point B? At face value, the different types of drive technology used in cars may be spectacularly different, but deep down, on a more basic level, what they have in common is the need to travel – mobility, to cover distance.

And thus, we introduce the following analytical distinction: We name something new, as seen by the producer, a novelty through diffusion: This is defined by the fact that existing objects are interpreted and designed differently. A distinction could be made between novelties – for example a product innovation – with the following:

Global novelty	A new product that is completely new for the global market.
New product range	A new product that enables a firm to gain access to a market that already exists.
Product range extension	A new product that complements an established product range of the firm.
Improved or re-developed products	A new product that is more effective or offers a broader range of benefits.
Repositioned products	An existing product offered in a new market/segment.
More cost-effective products	A new product, that deliver comparable performance at a lower cost.

Table 1: Differentiation of "novelties through diffusion", drawing on the example of product innovations

We name something new, as seen by the recipient, a novelty through adoption: This is defined by the fact that the need underlying the object is interpreted in a different way. Thus, whereas, for example, alternative drive systems are a novelty through diffusion, a novelty through adoption is, for example, an alternative mobility (travel) concept such as car sharing (car clubs), integrated concepts covering different modes of travel (e.g., offering the optimum combination of travel by train, car or bicycle from a single source). An example of a novelty through adoption could be:

Underlying Need	Old solution to the problem	Novel solution to the problem
Provision of food and drink	Supermarkets have to offer a broad selection	Aldi: Reduced assortment of goods, own-label, plain shops
Information	Information costs money	Google: Exchange information for information. Tell me what interests you and I will take you to interesting content.
Dining out	A restaurant is unique. A restaurant is marked by its atmosphere and the skills of the chef, who is a culinary artist.	McDonalds: Food is prepared by assistants who share tasks. Self-service. McDonalds is McDonalds, wherever you go.
Airline travel or travel over major distances	Flying is expensive and a luxury	Southwest Airlines: Reduce price by doing without luxuries, emphasizing standard airliners, packing seats in more tightly, and getting customers to pay for every extra.

>>

Cosmetics	Industrial, globally available cosmetics require animal testing and glossy advertising with models.	The Body Shop: The Body Shop completely renounces animal testing. The Body Shop promoted itself by campaigning against whale hunting, child exploitation, domestic violence and animal testing.
Listening to individually selected music	To listen to "my" music, I have to own records with "my" music on. Also, I have to do the work and find new bands that play "my" music.	Providers of music on demand: Subscription of transmission of music files via the Internet. The use of playlists that match my preferences and mood.

Table 2: Example of radical innovation in keeping with novelty through adoption[6]

A novelty through diffusion replaces previous "things"; a novelty through adoption replaces previous "concepts." Seen this way, it is possible to make distinctions within the phenomenon of "the New" based on whether the New is tantamount to changing the method used to solve a problem or a change in the solution to the problem. A first distinction of the New thus arises from the question: For whom is something new?

A further distinction which is derived from this lies in the issue of the degree to which an object is new or novel, that is: How new is something? Is it "a bit new" or "substantially/significantly/tremendously new?" Schumpeter also postulated that the "doing of new things" is "creative destruction" (Schumpeter 1946/1993: 136 ff.). Innovating thus means the destruction of the existing by actively spawning something that is essentially the New. The "creative destruction" Schumpeter had in mind manifests itself in revolution, in upheaval right down to the very roots by creating radical innovations. Naturally, the New can come about on a much smaller scale by renewing the existing; i.e.,

6 The named products and firms are partly protected brand names and property of the corresponding companies.

through incremental innovations. Naturally, although changes to the existing come across as less radical than replacing something with the New, incremental innovation can indeed still be destructive, albeit less through direct destruction and more through permutation. An initial analytical distinction can thus be made between innovations based on the (radically) New and innovations based on changes to the existing. In other words: Radical innovations come across as creative destruction; incremental innovations come across as creative (frequently less "dramatic" in terms of impact) change.

As plausible as this distinction between the New – as a radical or incremental innovation – may seem, it is still difficult to know how to apply these absolute distinction criteria. When is something significantly new or just a little bit new? Is a steam engine a radical innovation or just an incremental innovation, whereby, compared to a mill wheel, the power of water is transformed into mechanical energy in a different way, namely through another phase of aggregation? Is a railroad a radical innovation or just an incremental innovation, in which a steam engine is just used to generate kinetic energy mechanically? Is the automobile a radical innovation or just an incremental innovation, in which a machine-driven mode of travel has now just been shifted from tracks onto the highway?

The difficulty with making a distinction between the New in its incremental or radical form becomes more pronounced when one adds the distinction outlined above between a novelty through diffusion and a novelty through adoption. For example, in general terms, an incremental product innovation comes about when an existing product that has been launched is then adapted in esthetic, functional or technical terms without changing the essential core product and with this the product portfolio. Examples of a small, evolutionary change in the automotive industry are visible adaptations made to an existing vehicle to react to new fashion expectations.[7] At this level, things are still comprehensible and meaningful but the question remains: At what point does one change the essence of a product, or what actually is the essence of a product? Exaggerating to make a point: The essence of a two-axle carriage lies in the fact that it has four wheels, that it travels and that it is used to transport people and goods on highways and tracks. Does the type of drive change something

7 Sometimes such changes are referred to as "variations" and not as "incremental innovations." As already discussed in detail at the beginning of this chapter, there is naturally nothing wrong with the term "variation" – however, one must also accept that others might see this as an incremental innovation.

about the essence of a carriage? That is, does something about the very substance change if the carriage is now propelled by power supplied by horses, a combustion engine or fuel cells? Asking this a different way: Naturally, the radical switch from horse-drawn carriages to the automobile had a disruptive impact on companies, sectors within the economy and economies as a whole – but was this switch for users (at that time) of this vehicle so drastic?

> *Maslow's hierarchy of needs is still valid, and thus, at the very least, basic needs scarcely change. This is also reflected in products: What remains the same is the need for people to remove hair or clean teeth, and in their essence, the corresponding products have stayed the same. What has changed is the form of products with which the basic functions are carried out. For example, there are now electric toothbrushes which still essentially clean teeth the same way, identical to the older non-electric models with bristles, even if they are more efficient and quicker. The same applies to the Internet, that "merely" matches the information needs of humans, only in a dissimilar, dynamic way. Ultimately, at their core, these new types of products reflect the drive of companies in the market to differentiate themselves, expressed through the application of advanced technology to address an unchanging need – and this is not to be mistaken for fundamental change. (Translation, Freund 2013: 10)*

Looking at this the other way, it is also valid to say that with a novelty through adoption it is difficult to make distinctions between incremental and radical change. For example, the travel models above – car sharing; integrated multi-mode travel concepts – would be the incremental continuation of the concept of "covering distance horizontally." What could probably be considered a radical novelty through adoption in this area would be to translate the need to travel from the horizontal to the vertical: What if we could find everything we needed in life (a home, education, entertainment, shopping, services, etc.) in one and the same high-rise building? The actual need – to cover a distance that currently involves travelling from point A to B – would no longer be a challenge for car makers to master, but for elevator manufacturers. But is this radical novelty through adoption, that one could spend the whole of one's life in one and the same place, actually also essentially new or just an incremental development of something that was entirely normal and took place all the time in villages?

To ensure this is not misunderstood: We are not denying that these so-called radical innovations have a fundamental influence on society, and in no way are we denying the ingenuity of the innovators who came up with the ideas. But depending on how one looks at things, it is difficult – if not impossible – to distinguish between the fundamentally new and a gradual change in the existing. It is in the nature of innovations that they engender something new in the world. The first question regarding an innovation – one that is often not posed in the literature – is: For whom is something new? This has, at times, provided completely different answers to the question regarding the degree to which an object is new or novel, i.e., the answers to the following questions: "Is something new or not?"; "Is something radically new or an incremental change?" These follow no stringent logic; they are contingent, i.e., different observers will at times perceive one and the same object completely differently. It should also not be overlooked that the point at which something has not just been optimized, but could actually be considered new, depends entirely on the context within a sector of industry. In market segments in which the product life cycle and thus the general frequency of innovation works more slowly, this point is probably reached much more easily than in industries with shorter half-lives such as high-tech products and telecommunications.

The social context of innovation 2.2

In his book The Innovator's Dilemma (1997/2011), Clayton Christensen makes the distinction between the evolutionary and disruptive impacts of innovations. The distinguishing criteria between the two innovation impacts stems from whether an innovation focuses on the performance requirements or benchmarks on the supply side or on the demand side. Innovations with an evolutionary impact are characterized by the fact that something existing is continuously improved in keeping with the performance requirements of the supplier and the customer. So, for example, a product innovation with an evolutionary impact works like this: A firm equips an existing product with increasingly superior technology and new functions; this results in greater value for the customer with respect to the fundamental function of an object, e.g., enhanced performance, less consumption, simpler handling, reliability, etc. (Christensen 1997/2011: 6) Accordingly, every incremental innovation is also always an innovation with an evolutionary impact. Despite this, radical innovations can also have an evolutionary impact: An innovation would also have an evolutionary impact if, for example, a product is made which needs no more maintenance. This would result in no more need for classic business models relating to selling and supplementary services – certainly a revolutionary step for companies and a radical step. (cf. Matzler and von Eichen 2012: 59)

Innovations with a disruptive impact are fundamentally different in terms of the emphasis on expected performance requirements. A feature of such innovations is that they do not focus on the requirements or benchmarks of existing products, services, processes, etc. Instead, they completely reinterpret the fundamental function, i.e., the core action, the core task, the core effect of an object. The following shows examples of innovations with such a disruptive impact (see also Matzler and von Eichen 2012: 55-56):

Fundamental function of an object The "core function" of products, services, processes	**Innovations with an evolutionary impact** Products, services, processes, etc., which, gauged by performance requirements, have continuously	**Innovations with a disruptive impact** New products, services, processes, etc., which perform the fundamental function in a completely different way
Transport people and goods over water	Sailboat	Steam boat
Removal of earth, etc.	Cable excavator	Hydraulic excavator
Store in which all kinds of goods are sold	Department store	Discount supermarket chain
Measure and ascertain time	Mechanical watch	Quartz crystal watch
Clear and comprehensive presentation of entire existing knowledge covering all disciplines or only one specific area of specialty in alphabetical or systematic order	Encyclopedia	Wikipedia

>>

Equipment for spea-king to people far away	Landline telephone	Cell phone
Equipment for taking pictures	Camera	Digital camera
Transportation of people or goods through the air	Classic airline	Low-cost airline
Storage and consump-tion of sounds and audible media	Music CD	MP3
Information on current event	Printed newspapers	Electronic newspapers
Booking of vacations and travel	Travel agent	Online booking systems
Electronic data pro-cessing, enter-tain-ment, etc.	Cell phone, laptop, mobile Game consoles	Smartphone and tablet computers

Table 3: Examples of disruptive innovations and the products, services,processes, etc. they replace

2.3 Types of innovation

The differences between various "types of innovation" depend on the field or area in which a company or an organization is innovating. In the classic sense, innovations used to be and still are restricted to product innovation. So naturally it also comes as no surprise that whenever people referred to or still refer to research and development departments, what is meant is the area of a manufacturing enterprise that researches and develops new or changed products. However, concepts of "the New" and "the Changed" cannot and should not just be restricted to the world of business – they should also be part of and shared by civic society and the state. For example, what else could things like the green movement, life partnerships, social insurance and new teaching methods be called but innovations? As studies (e.g., Collins 2001: 162) and successful firms have shown, it is not necessarily new technology that help businesses steal a march on their competitors.

> *Innovations are no longer restricted to R&D laboratories and to published scientific papers [...]. New or significantly improved product, processes and methods in the provision of services; in business and organizational models; in low-tech industries; through creative imitation and technological catch-up; at the public level or at the level of society, all constitute innovations. (Dutta 2011b: 1)*

> *Innovations are not just of a technical nature: Services, organizational methods or processes can be innovative and have the objective of creating something new or making something better. (Translation, BDI&DTS 2011: 19)*

At this point, we return to the definition arrived at by Baregheh, Rowley and Sambrook (2009) after pulling together the content analysis of 60 definitions of innovation. They describe innovation as a process in which "organizations transform ideas into new/improved products, services or processes" (ibid: 1334). It is immediately apparent that the term "innovation" has been extended to the entire field of services. This is, of course, a tribute to the service industry, which has become a pillar of strength in the economies of "industrial nations." Also, this has something to do with the fact that, especially in mature markets, many products have almost no other way to differentiate themselves from one another than by offering supplementary support, i.e., through services. In the broadest sense, "processes" are also intended to mean that innovations are not just restricted to something provided or the results that are achieved for an external customer. Instead, "processes" is supposed to express the fact that innovations can also be intended to change something like production itself, or core activities (e.g., marketing, HR or finance), or even the way in which value is delivered to the customer (e.g., bricks-and-mortar trading versus online commerce, retailing versus direct selling, etc.). In this context, it is remarkable that the term "quality" has, for the same reasons, undergone similar developments in recent decades and that it now practically goes without saying that quality can be used to refer not just to products but also to services and processes.

Our perception of the term "processes" is that they are somewhat abstract; they express many things but at the same time nothing. If the same degree of abstraction were applied to "products and services," they would simply be referred to as "output." For a differentiated consideration of the various types of innovation (i.e., to look at the contextual dimension of innovations), in the following, we thus introduce the two most influential and best known authorities on this issue. These are:

> The Oslo Manual of the OECD ("Guidelines for collecting and interpreting innovation data")
> J. A. Schumpeter

Developed by the OECD, the Oslo Manual has pinpointed innovations of the following types: product and service innovations, process innovations, marketing innovations and organizational innovations.

A product innovation is the introduction of a good or service that is new or significantly improved with respect to its characteristics or intended uses. This includes significant improvements in technical specifications, components and materials, incorporated software, user friendliness or other functional characteristics. [...]

A process innovation is the implementation of a new or significantly improved production or delivery method. This includes significant changes in techniques, equipment and/or software. [...]

A marketing innovation is the implementation of a new marketing method involving significant changes in product design or packaging, product placement, product promotion or pricing. [...]

An organization innovation is the implementation of a new organizational method in the firm's business practices, workplace organization or external relations. (OECD 2005: 48-51)

The Oslo Manual simultaneously reduces and expands the term "process innovation." On the one hand, a process innovation is reduced to the level of production and logistics. On the other, the processes or activities of marketing, and the processes or developments within organizations as a whole are referred to as independent categories.

Schumpeter points out that (radical) innovation can be achieved as follows:

(1) The introduction of a new good – that is one with which the consumers are not yet familiar – or a new quality of a good. (2) The introduction of a new method of production, that is one not yet tested by experience in the branch of manufacture concerned, which need by no means be founded upon a discovery scientifically new, and can also exist in a new way of handling a commodity commercially. (3) The opening of a new market, that is a market into which the particular branch of manufacture of the country in question has not previously entered, whether or not this market existed before. (4) The conquest of a new source of supply of raw materials or half-manufactured goods, again irrespective of whether this source already exists or whether it has first to be created. (5) The carrying out of the new organisation of any industry, like the creation of a monopoly position (for example trough trustification) or the breaking up of a monopoly position. (Schumpeter 1934: 65 ff.)

In comparison to the Oslo Manual, it is immediately apparent that there is an additional type: access to a new supply market. It is also noticeable that the innovation types in the Oslo Manual are more of an operative nature, whereas with Schumpeter they are more of a strategic nature. With the former, a "marketing innovation" refers more to renewals of the marketing mix, whereas Schumpeter talks about entering a new sales market. Whereas the Oslo Manual talks more about an "organization innovation" to refer to organizational developments, with Schumpeter, it is about the strategic positioning versus the competition. From a modern perspective, one should make the following addition to the original Schumpeter definition (see also Faix 2008): Applied to today's thinking, the factor "introduction of a new method of production" can be extended with the introduction of new business processes, and the factor "new organization of any industry" can be replaced by "development and introduction of new organizational structures." The factor "conquest of a new source of supply of raw materials or half-manufactured goods" also relates to the "setting up of new (international) suppliers to access a new source of raw materials and semifinished products." According to Schumpeter, as far as companies are concerned, innovations can now fall into the following typologies:

- The development and introduction of a new product and the development and provision of a new service
- The development and introduction of new production methods and the development and introduction of new business processes
- Opening of new sales markets
- Opening of new supply markets
- The development and introduction of new organizational structures

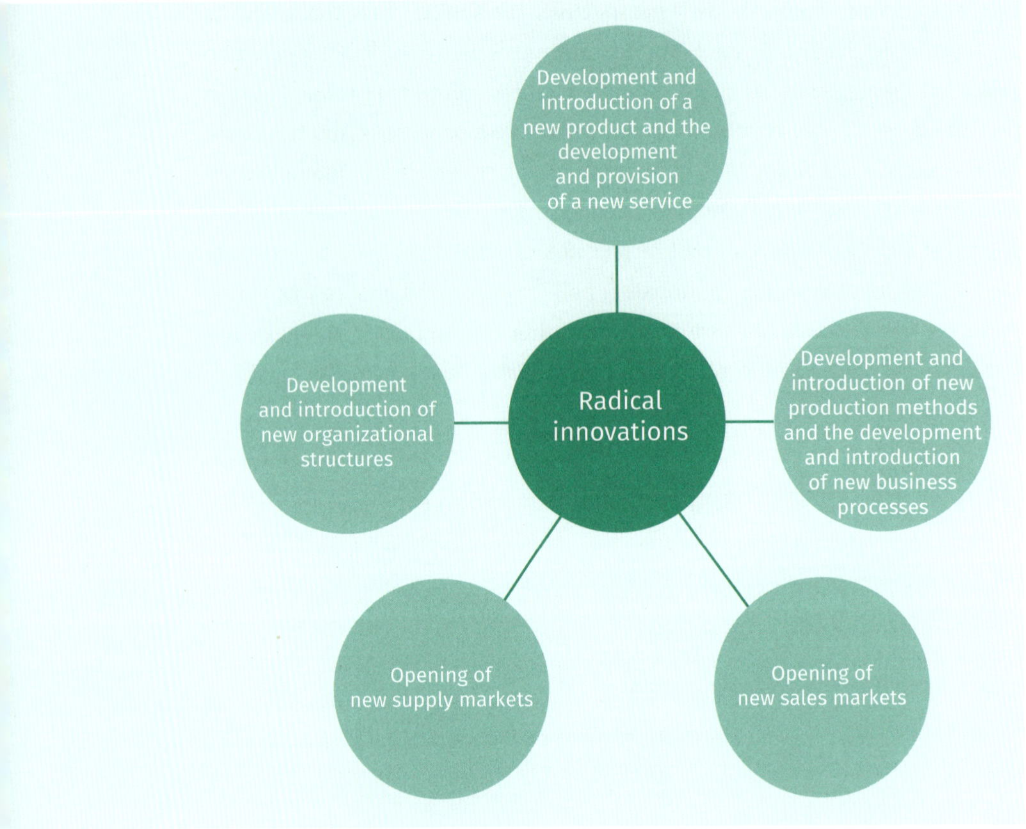

Figure 16: Schumpeter's types of innovation from a modern viewpoint (Faix 2008)

Aside from the typologies developed by the OECD and Schumpeter, there are of course myriad further attempts to place the "world of the new" into categories. Further, as time goes by, more and more innovations are made to innovations themselves. These initially relate to changes of a personnel nature within an organization – rules that are different from familiar rules governing tasks and processes. This includes areas such as the introduction of work on production lines or the principles of gemba kaizen – continuous improvement processes at the workplace. One could also point directly to the important area of financial innovation (the organization of capital markets, payment methods, options and sources of corporate funding, crowd investment, etc.). We have ascertained that such financial innovations are related to the real economy, so they are about the financing of and investment in enterprises and business activities. These are not innovations of the banking industry and they certainly have nothing to do with abstract objects of speculation and business models of casino capitalism, which have severely mangled the system of the classic market economy (naked sales of stocks, banks that operate without any equity and thus no liability). Furthermore, given the fundamental importance of the Internet, it is also worth mentioning infrastructure innovation (transportation, communication, etc.). Finally, it is also possible to use the term innovation if a business model is changed, affecting the following key business factors: changes to the value provided to certain customer or supplier groups; changes to parts of the internal or external structuring of service provision; changes to the selection or the mixture of sources from which the revenues of a business model are generated.[8]

One could thus add the following types of innovation:

> The development and introduction of new rules for tasks and processes
> The development and introduction of new investment and financial instruments, as well as procedures applicable to market players, that were previously not available on the money, credit and capital markets
> The development and introduction of new instruments and procedures relating to the movement of people, goods and information on the one hand, and access to goods and information logistics on the other
> The development and introduction of new business models (value proposition/value constellation/revenue models)

8 An example of an innovator in terms of value proposition is the online auction house eBay. The value proposition of eBay's business model is to provide liquidity for all kinds of exchangeable services. (cf. Stähler 2002: 79 ff.)
An example of an innovator in terms of the constellation of value creation is Dell, the second-biggest producer of PCs. Dell's business model is different from the traditional model of the PC industry in two ways. Firstly, Dell only sells PCs directly and does not include resellers, retailers and system integrators in the buying process. The second business model adaptation relates to the methods used to coordinate PC production. Under the classic model, production is based on sales forecasts with stocks of approx. 90 days. Instead, Dell only produces when it receives an order. (cf. ibid: 80 ff.)
An example of an innovation in terms of the revenue model can be seen with the prepaid pricing models offered by a variety of telecommunication companies. Instead of charging a flat fee for the telephone connection and issuing invoices after certain periods, customers pay an agreed amount in advance which is gradually used up on each telephone call. (cf. ibid: 84 ff.)

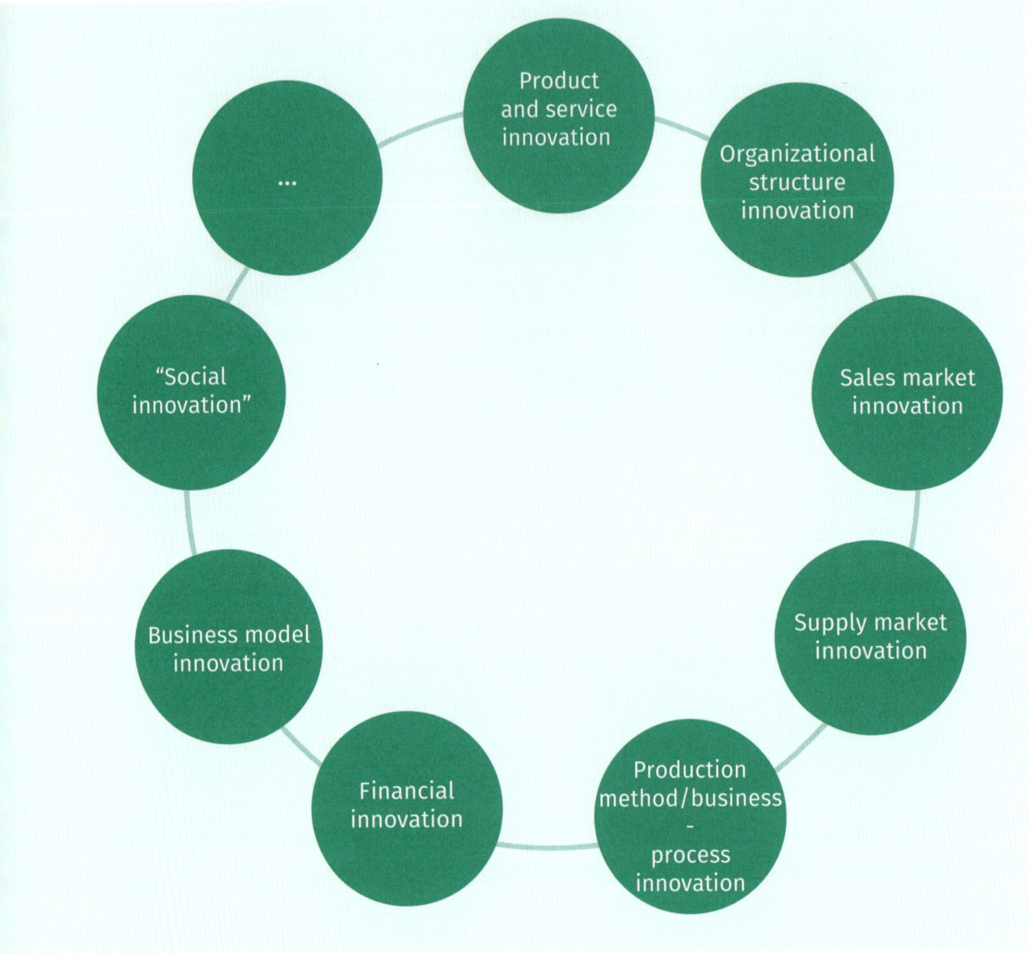

Figure 17: Types of innovation

Re-emphasizing the earlier point once again: When one uses the term innovation, one is not making a mathematical postulation, and, as a result, different points of view are not the exception but the rule. Our intention is therefore to add new impetus and to think as broadly as possible about the term InnovationQuality. As part of this, on the one hand people may argue about our innovation typology. On the other, this typology is not conclusive or complete; it should even be hoped that it will change. The very nature of innovation – precisely when it is radical and disruptive – is that it adds something to the world that did not previously exist in this way. We believe that this also applies to the term "innovation" itself, in other words, from time to time there will also be an innovation affecting innovations and new or changed types of innovation will appear. For example, in recent years we have witnessed business model innovation – i.e., innovations that have changed the "DNA of organizations" to a certain extent. Also, there have been system innovations whereby various preexisting types have been merged in a new way such that they represent much more than the sum of all underlying individual innovations. The intention behind this statement is: This typology, relating to the location and content of innovation is "work in progress" and not cast in stone – we also have to be open to possible surprises, otherwise we would actually be denying the original character of innovation.

Innovation – such a glamorous phenomenon 2.4

The "nature of innovation" refers to the deep, underlying essence of the phenomenon we describe as "innovation." One could also describe the "nature of innovation" and corresponding criteria of differentiation as a taxonomy of the New. In contrast to this, one could also refer to the different "types of innovation" as a nomenclature of the New. This analogy with biology appears highly appropriate because, as the social context shows, with innovation, one is just as likely to talk of evolution, emergence, extinction, recombination of possibilities, etc.

Furthermore, classifying nature has the following in common with classifying innovation:

1. The criteria of differentiation in taxonomy are more or less plausible, but not conclusive; answers to the question regarding what life or a living organism actually are, are just as much a working definition as answers to the question regarding what "the New" or an innovation might be.

2. The title and descriptions of the nomenclature are basically arbitrary and change constantly. What one calls a certain type of "the New," or whether one actually calls this new entity an innovation, are contingent decisions. Similarly, it is remarkable that suddenly some new kinds of "the New" may emerge and people launch into a discussion about whether this new entity should be adopted within the classification and what name it should be given.

In the broadest sense, we initially describe this phenomenon of "innovation" as the realization of something new. In doing so, it does not matter if the New is something incremental or radically new. It also does not matter if the New is new for its originator or the recipient (novelty through diffusion or adoption). It also does not matter if the New has an evolutionary or a disruptive impact. And finally, it does not matter if the New is a new engineering technique, a new technology, a new type of organization, a new way of doing business or something else which is new.

Step 1: Introduction to the two terms of "Quality" and "Innovation" by closely examining the phenomena and their distinctive characteristics

Step 2:

Combination of the two terms, or the two phenomena of "innovation" and "quality"

Step 3: Specific definition of the interpretation of "InnovationQuality", also as a quantitative variable

Step 4: Outline of a model for InnovationQuality

Step 5: Illustration of an approach for managing InnovationQuality

First attempt: Combination of the terms "innovation" and "quality" 1

In the following chapter, we describe possible starting points for combining the term "innovation" with "quality" – as well as the difficulties this creates.

The difficulty with interpreting and evaluating the New 1.1

The "creative destruction" Schumpeter had in mind manifests itself in revolution, in upheaval right down to the very roots by creating radical innovations. As an analysis of the Fortune Global 500 shows (Raisch, Probst, Gomez 2007: 46 ff.), sustained profitable success also requires complementary objectives, namely incremental innovation, i.e., continual reformation of the existing. It seems reasonable to determine what InnovationQuality is by making a distinction between radical and incremental innovation. For a start, "radical" and "incremental" are still not absolute entities but differences in the eye of the observer, based on a deeply individual continuum. Further, to enjoy sustainable success, companies have to yo-yo between the two supposed poles of "radical and incremental." Neither of these two poles is, per se, more valuable than the other: Radical innovations make a significant contribution to securing and expanding the future standing of a company; generally, incremental innovations account for the lion's share of financial success, i.e., for actually "harvesting the crops." Furthermore, it should be noted that it is not a distinction between radical and incremental that dictates whether a company is successful, but the customer, or in a broader sense, the user/party affected. And to clarify once again: The semantic differentiation into opposite poles of radical and incremental has a certain strategic role to play; yet – neither to realize company objectives, nor to deliver customer value – it is important not to prefer one pole over the other and then consider this the "measure of quality."

The New does not just appear like an offering from some transcendental location or a place beyond our planet. Both incremental and radical innovations are based on the principle that existing entities are combined in a new or a not yet implemented way, or as Schumpeter also said: An innovation is the result of successfully combining factors (e.g., products or services, ways to source or sell) in a new way. (cf. Schumpeter 1934: 65-66) With supposedly incremental, but also with supposedly radical innovations, this concept is of course different in a "gradual" sense, but not as a matter of principle. The "gradual" difference lies in what is combined with what. At this point, let us take an example "The horse-drawn carriage versus the automobile":

For hundreds of years, the concept of the horse-drawn carriage was the dominant solution for the problem of covering distance over land. As the centuries went by, time and gain the concept was changed incrementally. For example, there was "product variation" in the number of horses (one, two, four, etc.) or the type of horse that pulled the carriage (draft horses, heavy warmbloods, etc.). Aside from the desire to make a statement, there were practical reasons for the number and the type of horse, depending on the load that needed to be transported or the speed of the journey. The number and the type of horses is an essential and (at least as far as the number is concerned) obvious factor when it comes to the performance of a horse-drawn carriage. The equation behind the incremental innovation of the horse-drawn carriage was thus: combine carriage with a certain number of horses or a certain type of horse. Naturally, Carl Benz saw the equation behind the radical innovation of the automobile completely differently: combine carriage with a combustion engine. However, in a manner of speaking, this kept an essential and obvious criterion of differentiation for performance "under the hood." Although machines and engines had of course already been in existence for some time in those days, for the majority of the population, they were less well known in public – unlike horses, which were a public, everyday feature of life. Presumably, a significant proportion of the population could thus judge the performance of a horse or a breed of horses. In contrast to this, probably only a very small section of society was in a position to judge the performance of a (combustion) engine. Even if many could see the potential of the automobile innovation, a significant number of their contemporaries mocked and derided the "wagon without horses."[9] At this point, one thing is totally apparent: Evaluation of the "radical new" depends to a large extent on whether or not, and to what degree, the subjects making the evaluation are themselves immersed in incremental thinking.

9 It is worth remembering the words of the last German Emperor, William II: "My money's on the horse – the automobile is just a passing fad."

With a radical innovation, the originator and the recipient of the innovation are entering completely new territory. This also means that – as the originator or as the recipient – one has to come to terms with completely different performance requirements and quality benchmarks compared to those that were established for previous products or previous industries. Naturally, some highly significant characteristics differentiate a horse from a combustion engine. To start, it may make sense to choose simple benchmarks like the size, or which is the "leader of the pack" in each field that each of the objects comes from (for example, among carriage manufacturers and car makers) and combine this.[10] Nonetheless, it should be noted that a radical innovation does involve combining things in a way not done previously. Or expressed in more simple terms: The result of such a combination is more than the sum of all parts. Subsequently, a company that is creating something radically new has no choice but to set itself targets, e.g., regarding the nature of performance and the outcome that will be achieved.

Given all the difficulties presented by radical innovations, ultimately one consequence in terms of the quality of innovation would be to focus solely on incremental innovation. Initially, it should be pointed out that the dividing line between incremental and radical can neither be drawn precisely through analysis, nor can one or the other be an advantage in a business sense. Furthermore, naturally such a focus would not only severely restrict the areas that the term "InnovationQuality" could be applied to, for many, this would also make the term obsolete. In many areas of innovation research, as well as in the public at large, radical innovations are also seen as "innovations in themselves," as innovations in the truest sense. This may primarily be because of the nature of the essentially new – after all, it often results in important and/or pressing problems being solved.

Secondly, this may be due to the nature of the kind of people who are interested in the essentially new. Thirdly, it may be because the advent of the essentially new can result in the emergence of new companies and sometimes even new branches of industry.

Implementing the New is the very essence of what one means by innovation. The first difficulty when it comes to the quality of an innovation lies in the fact that the New can manifest itself in a variety of ways. One general and abs-

10 It is little wonder that the performance of engines is still given in "horse power."

tract differentiation is the distinction between a novelty through diffusion and a novelty through adoption. This stems from the question as to who decides or perceives whether something is new. The two fundamental answers to this are: the originator or the recipient. At this point, it is clear that the two phenomena "innovation" and "quality" have something remarkable in common because, as with quality, a distinction can be made between who decides or perceives whether and to what extent something is of good quality. In the literature, a distinction is made between producer-based and customer-based quality. A stated objective and the aspiration of this book is to define and outline a general term called "InnovationQuality." As a result, it is necessary with such a term to subsume all aspects and all angles. In other words, for the term "InnovationQuality" to be comprehensive, it has to include the "radical new" and the "incremental new" and be applicable to a novelty through diffusion and a novelty through adoption, thus relating not just to the originator but also to the recipient.

1.2 When quality becomes a disaster

A central insight in Clayton´s book The Innovator´s Dilemma is that radical innovation – and thus innovations with a disruptive impact – are sometimes drastically inferior to existing objects on key performance criteria: For a long time, the sound quality of MP3 was significantly inferior to CDs; the same applies to the image quality of digital cameras compared to conventional cameras with a film; the discount supermarket chains offer much poorer customer service and an inferior selection of products compared to traditional outlets, etc. When a disruptive innovation is introduced, there is only a small niche, ignored by the established "leaders of the pack."

But within this niche, such innovations are hugely successful and can keep developing, tweaking their performance criteria upward in an evolutionary manner until they do a better job at fulfilling the fundamental value of an object than existing solutions, ultimately replacing them.

When Robert Fulton was navigating the Hudson River in his steamboat in 1819, steamboats were inferior to sailboats in almost every respect: The costs per nautical mile were higher, the boats were slower and much more susceptible to problems. On a fundamental level, steamboats were completely unsuitable for ocean travel and could only gain a foothold in a completely different market. At first, their market was travel on inland waterways, where the benchmarks of performance where totally different. On rivers and lakes, it was periodically necessary to travel into the wind or when there was no wind. And it was precisely here that steamboats were superior to sailboats. The actual problem was not about knowledge regarding the new technology used in steamboats. The problem was more that the producers of sailboats listened to their customers. For a long time, the boatyards could not use steamboats for ocean travel. The first steamboats were slow and unreliable. They needed support from sails. It was not until 1889 that the first ocean-going steamboat entered service without any sails. The steamboat traveled at 20 knots, now making it a contender for the sailboat. To also succeed in the steamboat market, the producers of sailboats would have had to focus on the inland waterway market because this was the only market in which steamboats were of value at the time. But what did they do? They ignored the new technology and focused on further development of the sailboat – and thus on the much bigger and more attractive market. Step by step, the performance of the new technology improved. Soon, steamboats were just as reliable and fast as sailboats and thus no longer to be stopped. (Translation, Matzler and von Eichen 2012: 53–54)

According to Christensen, the Innovator's Dilemma (2011: 125 ff.) lies in the fact that companies that innovate on an evolutionary level are often not doing anything wrong – except that they tried to keep giving their customers something better.

Companies start losing their composure for a multitude of reasons: bureaucracy, arrogance, weak management, poor planning, short-termism with investments, insufficient skills and resources, but also simply bad luck – to name just some of the reasons. We observe time and again how leading companies get it wrong when it comes to a switch to groundbreaking technology in their industry. Our focus lies in the well managed companies. Companies that extend their antennae, meticulously analyze the needs of their customers, aggressively invest in innovation and the ongoing development of their technology and products, that have developed outstanding planning and decision-making processes, and have a marked focus on revenues and growth. These are characteristics we would ascribe to every successful company – the ultimate key success factors. This makes it all the more astonishing that it is precisely these factors that become a disaster for the established leaders of industry when they encounter a disruptive change in technology. They fail because they are customer-oriented, because they keep redeveloping their products and because they are focused on revenue and growth. (Translation, Matzler and von Eichen 2012: 51)

What this means when it comes to interpreting the term InnovationQuality is that, where disruptive innovations are concerned, one cannot simply transfer conventional expectations and "quality ingredients" to other areas. Further still, in so far as it is possible to have factors of quality and these can be considered design principles in a broader sense, these classic success factors can result directly in a company's downfall. Customer understanding, empathy, asking customers what they want, tapping into customers wishes or even second-guessing them, benchmarking, introducing best-practice processes, the optimizing and standardizing of production procedures used in existing processes, etc. – these are all the kinds of things you really should not do if you need/want to innovate disruptively! Once again, in clear terms: It is also, or precisely because evolutionarily innovative companies did many things "correctly" when it came to classic ways of interpreting quality, that they eventually foundered; and disruptively innovative companies have mainly been successful because – at least at the beginning – when it came to classic ways of interpreting quality, they consciously or subconsciously did things "incorrectly" (or at least differently). Using a variation of the well-known words of Voltaire that "perfect is the enemy of good": The best – as a substitution – is the victorious enemy of good – as an optimization.

Beyond new products 1.3

To express the different types of innovation in an extremely general way, one could make a distinction based on who the "recipient" of an innovation is or who is affected by it. In the broadest sense, one can make a distinction between externally oriented and internally oriented innovation.

One way to describe quality is as the result of an evaluation process. An object, in this case the innovation, is judged by a subject according to a defined yardstick. There are, however, striking differences between subjects who carry out an evaluation, especially with respect to externally oriented and internally oriented innovation. The evaluators of an externally oriented innovation are primarily customers, third-party auditing bodies, journalists, etc. The evaluators of an internally oriented innovation are primarily managers, other employees and business partners. Outsiders are only indirectly affected by internally oriented innovation such as the organizational restructuring of a company – e.g., measures for achieving synergies through organizational development, improved collaboration, etc., which are ultimately about improving or changing products and services. By implication, outsiders can provide feedback on the quality of this internally oriented innovation, but only indirectly, i.e., by pointing to some kind of direct impact. The immediate aim of many internally oriented innovations is to make the organization quicker, leaner, more agile, "smarter" or more efficient through renewal or change. An objective derived from this could be that the customer gains greater/better value stemming from the services provided or results. But another derived and often intentional aim with internally oriented innovation can also be that the customer should actually not notice anything. In other words, even if the internal workings of a firm should undergo spectacular change, the quality of the products and services it offers should, at least, remain the same. If no distinction were made at this point between externally and internally oriented innovation, this would clash with the actual definition of an innovation, namely, that something new has entered the world or an existing object is being changed. In a nutshell: At best, "outsiders" can only indirectly evaluate an internally oriented innovation – if at all.

This does of course also work the other way round: In an absolute and transcendent sense, quality is the subjective experience of a person relating to the nature of an object. Even if this totally obvious interpretation of quality is of little scientific relevance, since it cannot be expressed objectively (cf: rigidity versus relevance), once the overall picture comes together covering multiple categories, it provides the key reason for a subject to perceive something as good or bad. So that we are not misunderstood: Naturally, quality should – indeed must – be captured in a different way and in particular it has to be measured. The yardstick for this is provided by object-related and manufacturer-related key indicators for quality. Every producer and service provider must however possess the humility to accept that, on the one hand, quality as perceived by the customer is not just a question of the nature of the object, but also of a firm's image and marketing; on the other hand, the ultimate yardstick for the quality of a service delivered by a company is not a standard, but instead the criteria of the customer. Of course companies can determine the objective nature and the image they portray for an externally oriented innovation, but naturally such aspects are only correlative and not causal when it comes to the perception and experience of the customer. The ultimate evaluation of an externally oriented innovation is a profoundly opinionated matter, a self-determined action of the recipient. A company may be able to influence the answer to the question "What does the customer think of something?", but they cannot determine it. To exaggerate to make a point, this means that "insiders" can at best only indirectly evaluate externally oriented innovation, or – if at all – only as their own customer.

Insofar as one accepts any kind of typology, extending to the New beyond new products and services, there is a further problem: It goes without saying that product innovation is something completely different and works in a completely different way to sales market innovation, business model innovation or organizational innovation. The classic interpretation of quality leans more towards products and services; most literature relating to "quality" looks precisely at these two areas. But a general term for "InnovationQuality" must encompass innovations beyond the realms of new products and services.

Accepting that there are different types of innovation ultimately means that, in principle, innovation is not just possible in all parts of a company – in all departments, in all functions – but that it is also necessary. Initially, this banal insight was, in terms of quality, not translated into a reality or reflected in tangible action; it was only when people started to accept the broader interpretation of quality that it dawned on them that quality is not just the responsibility of the "quality department," but also the responsibility of all employees and all business functions. There is an essential commonality here with "innovation": Actively working on innovation – in keeping with the concept of total quality management (TQM) – everywhere in the company, with every employee involved; the "doing of new things" should not be limited to research and development departments.

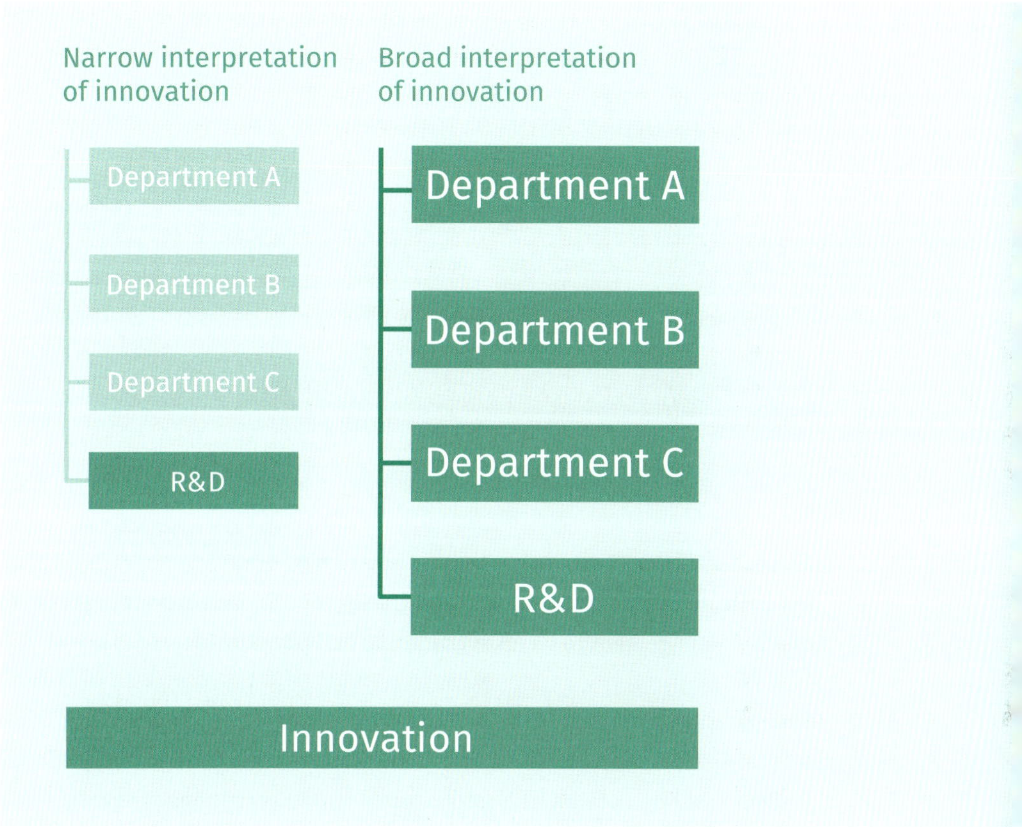

Figure 18: Innovation-oriented self-interpretation in all departments of a company

Furthermore, the quality of innovation is thus a decisive objective within a company and not just within the department which is "responsible for it."

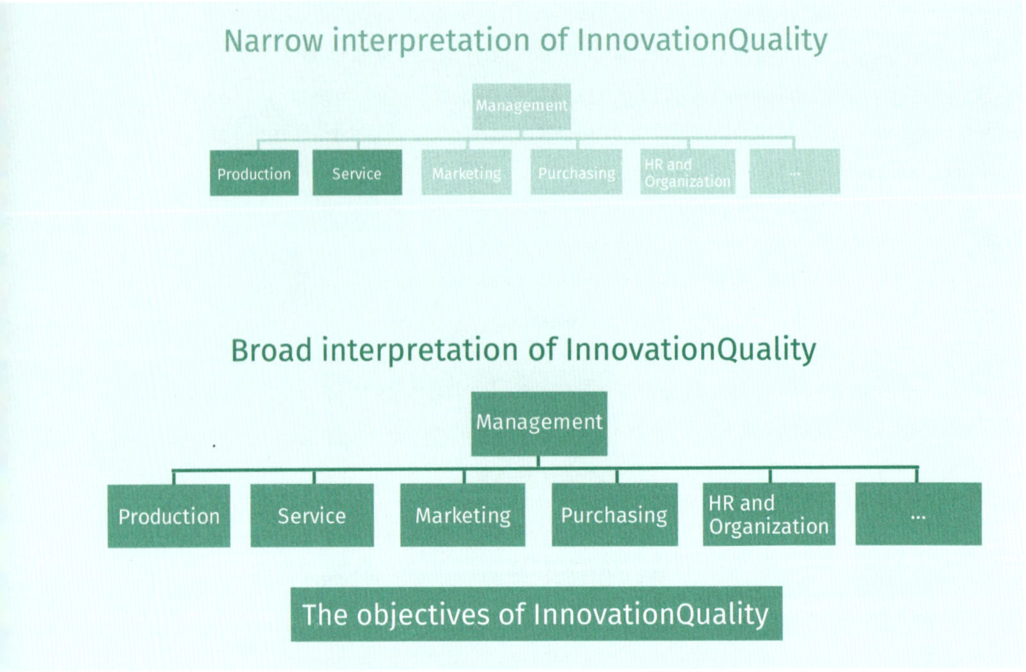

Figure 19: Innovation-oriented self-interpretation in all areas of a company

The general term "InnovationQuality" adds a new dimension to calls for quality to be an overarching objective and, with this, for corresponding tasks of quality management to be an overarching activity within a company.

Summarizing these points: Even if the terms "innovation" and "quality" have several distinct characteristics, which can certainly result in fruitful synthesis, it should still be noted that classic concepts and perceptions of "quality" can scarcely be transferred to the phenomenon of "innovation" – if at all. With incremental innovations, the term "InnovationQuality" would be relatively easy to derive: Quality would entail fulfilling one's own performance standards, or those of the competition, or even exceeding these ("improvement innovation"), or meeting or exceeding expectations and standards within one's environment. For innovations with a radical or disruptive impact, completely different rules apply. And, stating this clearly once again: Continuously improving the quality of products and services, customer empathy and customer focus are all effective and recommendable success factors of quality management, but they may harbor the seeds of disaster when it comes to the phenomenon of "innovation" and indeed they can actually hamper innovation.

Second attempt: Derivation of a comprehensive, general and pragmatic interpretation of "InnovationQuality" 2

As is well known, according to ISO standard 9000: 2005 the definition of quality is: the degree to which a set of inherent characteristics fulfills requirements. In the following, we would like to attempt to transfer this definition to "innovation" one step at a time. Accordingly, the following can be defined to start with: InnovationQuality denotes the degree to which a set of inherent characteristics fulfills requirements.

The "degree" of InnovationQuality 2.1

One of the main features of quality is that there is an objective or subjective standard, a gauge by which, for better or for worse, we categorize "things." Of course, even if there are many elaborate ways to determine quality, quality is closely linked to the fact that, with objects, we make comparisons. Maybe we regard them with respect to other, real "things" which are similar in essence. Maybe we compare objects with an imaginary ideal. Qualitative judgments, for example "immaculate," "unblemished" or "impeccable," imply that we have some kind of expectation of what an object should be like, what an object should do, or what an object could be worth to us. "Innovation" – especially in its radical or disruptive form – fundamentally goes against the grain of this quality expectation: Innovation is simply sometimes the "essentially new," something that has never existed in this way before – or in a nutshell, something one had never imagined before in this way, let alone anything like it. The fundamental nature of an innovation is thus that it represents something incomparable. Quality and innovation are in their very essence of a completely different nature. And thus, if one compares distinct characteristics – in terms of the actual nature of innovation and quality – it is actually not possible to merge the terms innovation and quality. So in simple terms: When understood to mean the quality of a product or structure, the term "quality" can hardly, if at all, be transferred to the term "innovation."

More recent quality management models such as Six Sigma are known for the fact that their aim is to achieve as few deviations as possible or even "zero-error production." But this is also contradictory if one attempts to transfer this expectation to the term "innovation": An innovation is the result of a deviation from the existing, indeed in its very essence it can be nothing else. If no or only minimal deviations are permitted from the standard, and if such deviations are even referred to as "errors" to be eradicated, then this spells certain death for innovative work and efforts. Coming back once again to the earlier point: Quality means deviating as little as possible from a predefined standard. Innovation means, deviating as much as possible from a pre-defined standard. So in simple terms: When understood to mean the quality of production or a process, the term "quality" can hardly, if at all, be transferred to the term "innovation."

Summarizing the points raised until now, it should be noted that the term "quality," if interpreted in terms of structure and process quality, essentially revolves around laying down targets and control variables. With innovations, however, such quantitative target variables actually have a paradoxical effect: Quantitative targets and control variables are the result of a process of systematically observing, capturing and analyzing a past reality based on existing rules. The nature of innovation, however, is that, on the one hand, it focuses on the future; on the other, it is something essentially new that sometimes more or less breaks with the past. By attempting to capture innovation in terms of structures or processes, one misapprehends the actual nature of innovation: that innovations are inherently surprising; they not only break rules, but – in keeping with C.G. Jung – are, in essence exceptions to the rule.

It should also be noted that classic models – and thus also quality criteria – tend to be, or are entirely related to products and services, or production processes. So this also means that transferring such models to other types of innovation will at the very least be difficult. The term "innovation" encompasses a variety of highly different phenomena. And it appears to be difficult, if not impossible, to determine the degree of "InnovationQuality" by using objective and exact indicators which are simultaneously transferable to a broad variety of connections, some of which may not yet be known (cf, "innovation of an innovation").

So in brief, the following can be stated: Quite simply, InnovationQuality can not be expressed in terms of structure or process quality. Instead, "In-novationQuality" should be seen within the context of "the quality of the result" or "outcome quality": In our opinion, the term "innovation" can and should mean more than "new products." Instead, "innovation" should be interpreted from a business angle as everything new and changed that was cultivated by a company in the hope that it has a positive influence on the "life" of the company – for example, that the company feels healthier, more viable, more vital, etc. Aside from M&A activities, innovations are the essential cause of positive change in all entrepreneurial undertakings. And of course product innovations are a key reason for this change. To overstate the case, companies are there to solve the problems of their customers or satisfy their needs; to solve problems and satisfy needs, companies provide their customers with products and services. In short, the purpose of enterprises is to address the problems and needs of their customers with products and services. As a result, product and service innovations must naturally be highly valued. But there are scores of other ways for companies to positively influence the "life" of the company through the New and the changed. And thus InnovationQuality is a factor which makes it possible to show whether and to what extent the "life" of a company is being changed for the positive by an innovation. Whereas we initially described InnovationQuality as the degree to which a set of inherent characteristics fulfills requirements, we can now expand this into the following: InnovationQuality denotes the degree of an outcome, whereby a set of inherent characteristics fulfills requirements.

In the previous chapter, a three-dimensional, multi-factor construct of the third order was introduced for quality. Given this, when it comes to the term "InnovationQuality," it seems to us that the only way to take this further is with the following conceptual approach:

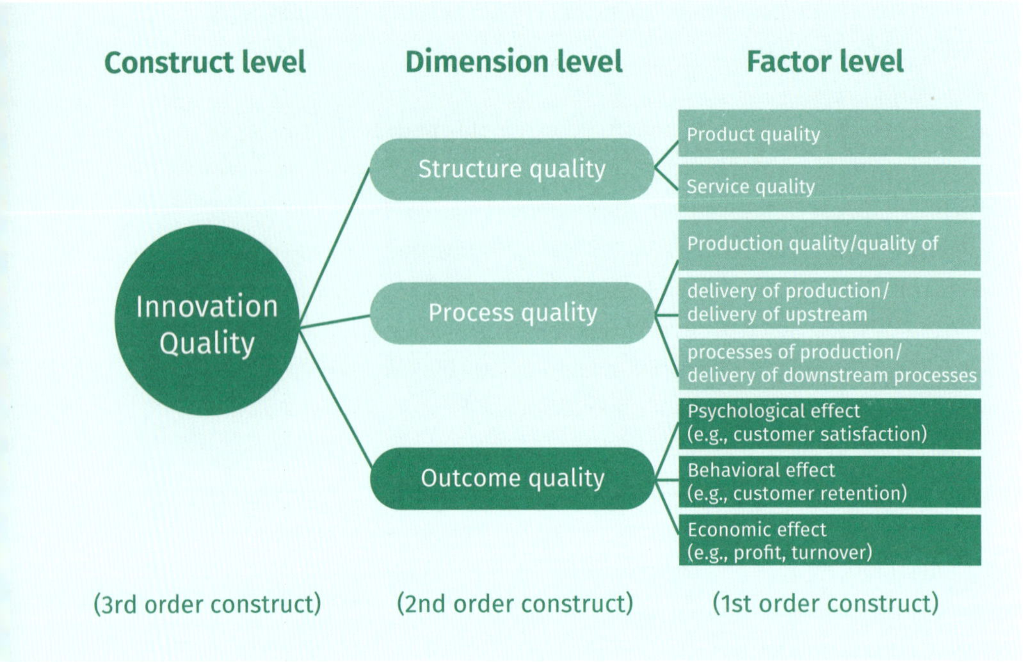

Figure 20: Definition of InnovationQuality as a three-dimensional, multi-factor construct of the third order

2.2 The "set of inherent characteristics" of an innovation

The predicator "set of inherent characteristics" is used to mean everything grouped together that can constitute an "innovation." As should be clear from the previous chapters, the "inherent characteristics" of the phenomenon of "innovation" are not at all easy to capture in a clear and distinct "set." We have already described the phenomenon of "innovation" in its broadest sense, as any realization of the New, and it does not matter if the New is something incrementally or radically new. It also does not matter whether the New is new for the originator or the recipient (novelty through diffusion vs. novelty through adoption). Nor does it matter whether the New has an evolutionary or a disruptive impact. Finally, it does not matter whether the New is a new engineering technique, a new technology, a new type of organization, a new way to do business, or something else which is new. And thus a distinction should only be made with innovation in general, in keeping with Schumpeter:

Innovation lies in doing something new or doing something that is already being done in a new way. It can thus be said that: InnovationQuality denotes the degree of an outcome, whereby an innovation fulfills requirements.

The "requirements" of an innovation 2.3

The last predicator that needs explaining is thus "requirements." The key vital operations of a company, i.e., the reason for and manifestation of entrepreneurial "life" is that a company creates value. The greater the value created, the greater the vitality of a company. And if a company operating in a market economy no longer creates value, this is tantamount to its demise. The requirement a company has of an innovation is thus, generally, that there is a positive influence on how value is created as a result of the innovation.

But what actually is the "positive influence" an innovation can have on entrepreneurial value creation? To examine this, one needs to return to the general purpose of an innovation: On a general level, an innovation results in a shakeup of the homeostasis, the equilibrium, or the synchrony of a system. Transferred to a company, this means: An innovation influences the existence and essence, the nature and form of value creation. The reason for the shakeup can be of a reactive or proactive nature. In other words: Either a specific change takes place on the outside (within the environment of the system) to which the system must react with a change on the inside; or the system anticipates a possible change in its environment and it preempts this with a change on the inside. On one hand, the aim of an innovation is thus to preserve viability in the future by introducing necessary changes to the existence and essence of the system, given the actual changes in the environment. On the other, the purpose of an innovation is to provide a system with a vital advantage compared to other systems by preempting necessary changes to the current existence and essence of the system, given possible changes in the environment.

In other words, and to make this clear: One purpose of innovation is to safeguard and maintain the viability of a company to create value (given changing conditions) by influencing the existence and essence of the company's own value creation. Another purpose of innovation is to obtain an advantage over other companies (also given changing conditions) by influencing the existence and essence of the company's own value creation.

In a nutshell: The purpose of an innovation for a company is to initiate, safeguard and sustain value creation. And thus it can be said for InnovationQuality in a broader sense that, InnovationQuality denotes the degree of an outcome, whereby an innovation initiates, safeguards and sustains value creation.

2.4 In summary: A definition of "InnovationQuality"

For a company, an innovation performs a fundamentally vital function, since an innovation makes it possible to initiate, sustain and safeguard value creation.

In its broadest sense, we describe the phenomenon of "innovation" as the value-creating realization of the New. It does not matter if the New is something incrementally new that sustains or safeguards current value creation, or whether it is something radically new that initiates new value creation. It also does not matter if the New is new for the originator or the recipient (novelty through diffusion or novelty through adoption). Nor does it matter whether the New has an evolutionary or a disruptive impact. Finally, it does not matter whether the New is a new engineering technique, a new technology, a new type of organization, a new way to do business or something else which is new. For the term "InnovationQuality," this thus means:

InnovationQuality denotes the degree to which there is a certain outcome, resulting from an innovation initiating, sustaining or safeguarding value creation.

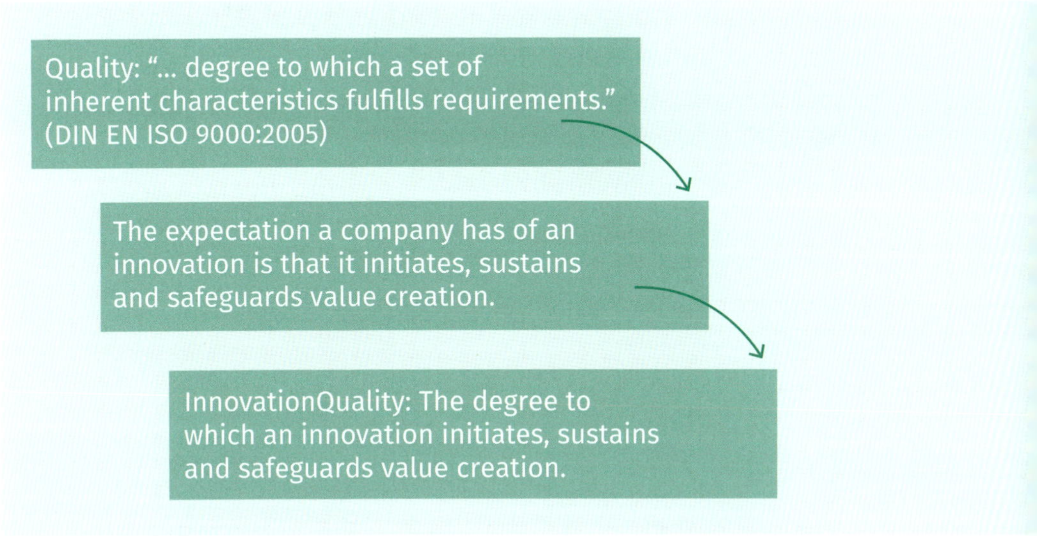

Figure 21: Quality and InnovationQuality

Step 1: Introduction to the two terms of "Quality" and "Innovation" by closely examining the phenomena and their distinctive characteristics

Step 2: Combination of the two terms, or the two phenomena of "innovation" and "quality"

Step 3:

Specific definition of the interpretation of "InnovationQuality", also as a quantitative variable

Step 4: Outline of a model for InnovationQuality

Step 5: Illustration of an approach for managing InnovationQuality

According to the definition derived in the previous step, InnovationQuality denotes the degree to which there is a certain outcome, resulting from an innovation initiating, sustaining or safeguarding value creation. Based on these definitions, the next step is to introduce and discuss quantitative variables, including formulae, to define how "InnovationQuality" is interpreted. We use "formula" in a general sense to mean a short, to-the-point and coherent representation of a connection, complete with suitable components, whereby we acknowledge the risk of over-simplification in order to pave the way for hopefully fruitful "discussion after reading" – something that is important to us.

The term value creation as a retrospective definition 1

The term "quality" has a fundamental connection with the term "value." While elaborate ways are found to talk about the indicators of "quality" in the world of science, the term "quality" is overloaded and overstretched in industry. For example, for some customers a "quality product" quite simply means that a product is "more valuable" than other products in one way or another. Such absolute interpretations of value always imply that an object or an action are of no value in themselves. Instead, the value of an object or an action are the result of a process in which a subject judges the value resulting from the availability (possession, property, etc.) of an object or from being involved in an action (consumption, etc.). In turn, the interpretation of value is based on the fact that the subject making the judgment is in a teleological dilemma before making the evaluation, i.e.: The subject must on the one hand have a need and on other hand have an expectation relating to a desired but not yet arisen state. The object or the action are evaluated as a means to an end (the purpose) in order to resolve this teleological dilemma. In summary: The terms *quality*, value, benefit and purpose are closely connected in meaning. The quality of an object or an action is dictated by the value a subject assigns to the object; in turn, this value is a result of the extent to which an object is of value to a subject. Or, expressed more succinctly: The value of an object or an action is dictated by whether and to what extent it may fulfill a purpose for the subject.

As is often the case, the whole is more than the sum of all parts, so putting together "value creation" is about more than just adding the terms "value" and "creation" to one another. In a monetary economy, value creation means transforming something that already exists – with a certain monetary value – into something else – with a higher monetary value. Expressed in more general terms, this means: Value creation is the outcome of a subject evaluating something that was created to be more valuable to him than it was before. Transferring this to the term innovation: Value creation through innovation is a reflection of whether and to what extent something new appears to be more valuable to a subject than something old. This relationship can be expressed in relative or absolute terms.

In relative terms, value creation resulting from innovation (abbreviated to $cVIn_{relative}$) can generally be defined as a quotient of the outcome after diffusion of an innovation over the outcome before diffusion, i.e., before the innovation was shared.

For this formula also to be valid for radical innovation – i.e., the introduction of something sometimes completely new – the outcome before diffusion of the innovation has to be at least 1; or looking at it the other way round, if the divisor is 1, this shows that an innovation is a radical innovation.

$$cVIn_{relative} = \frac{\text{Outcome after diffusion of an innovation}}{\text{Outcome before diffusion of an innovation}}$$

Formula 1: Value creation expressed by cVIn (relative)

As a rule, this means a quotient of "$cVIn \geq 1$" indicates that an innovation results in positive value creation; a quotient of "$cVIn = 1$" indicates that an innovation results in neutral value creation, i.e., value is retained; a quotient of "$cVIn \leq 1$" indicates that an innovation results in negative value creation, i.e., value is reduced.

$cVIn \geq 1$ means value creation
$cVIn = 1$ means value retention
$cVIn \leq 1$ means value reduction

Formula 2: Interpretation of the factor VIn

The value creation of an innovation in absolute terms (abbreviated to $cVIn_{absolute}$) is the result of subtracting the outcome before innovation diffusion from the outcome after innovation diffusion.

$$cVIn_{absolute} = \text{Outcome after diffusion of an innovation} - \text{Outcome before diffusion of an innovation}$$

Formula 3: Value creation expressed by cVIn (absolute)

The advantage of keeping a formula for the value creation of an innovation this general is, first of all, that it can be used in non-business contexts. Secondly, the formula can also be used in areas of business where non-monetary factors are more meaningful. For example, with the introduction of an organizational innovation, indicators such as "employee satisfaction," "employee fluctuation," "knowledge sharing," etc. are not just highly relevant, but also potential precursors of possible influences on a firm's financial results. Finally, within a business context, a formula kept general in this way would appear to make sense if a monetary result can only be arrived at in hypothetical terms by looking at causes and effects. For example, this is pertinent to the question as to whether and to what extent the impact of advertising or PR activities can be deduced from business results.

In principle, given the overall importance of InnovationQuality to a business, the task and aim should be to express any outcome in monetary terms, whether this is in terms of the contribution to profits or turnover, or in terms of the contribution made to saving money inside or outside the company (which of course would subsequently be the same as a contribution to profits). In systems theory terms, specific outcomes should be expressed through a general medium used in the economy: monetary value. Thus, in a business context, the following formulae can also be derived to express value creation resulting from an innovation:

$$cVIn^*_{relative} = \frac{\text{Turnover after diffusion of an innovation}}{\text{Turnover before diffusion of an innovation}}$$

Formula 4: cVIn (relative) within the context of the business indicator "turnover"

$$cVIn^{*}_{absolute} = \text{Turnover after diffusion of an innovtion} - \text{Turnover before diffusion of an innovation}$$

Formula 5: cVIn (absolute) within the context of the business indicator "turnover"

$$cVIn^{**}_{relative} = \frac{\text{Profit after diffusion of an innovation}}{\text{Profit before diffusion of an innovation}}$$

Formula 6: cVIn (relative) within the context of the business indicator "profit"

$$cVIn^{**}_{absolute} = \text{Profit after diffusion of an innovation} - \text{Profit before diffusion of an innovation}$$

Formula 7: cVIn (absolute) within the context of the business indicator "profit"

Similarly, a variety of business performance indicators can be used (key success indicators such as EBIT and the like, profitability indicators such as return on investment and such).

Another indicator that can provide extremely useful insights in connection with the impact or the outcome of innovation is called "Return on Innovation" (ROIn).[11] This indicator provides a point of reference for the cost and returns of an innovation. ROIn can be expressed in relative or absolute terms.[12] To express the return on innovation in relative terms (abbreviated to $ROIn_{relative}$), a quotient can be used based on the returns after diffusion of the innovation and the overall cost of making the innovation possible. To express the return on innovation in absolute terms (abbreviated to $ROIn_{absolute}$), the overall cost of making the innovation possible (e. g., concept development, licensing, prototypes, advertising, sales, etc.) is subtracted from the returns after diffusion of the innovation.

11 Cf. also the performance indicator "Return on Quality (ROQ)" (Rust, Zahorik and Keiningham 1994 or Kamiske 1996)

12 The indicator ROIn, or rather, ROIn in absolute terms was not invented by us; its originator and actual creator appears to be unknown.

$$ROIn_{absolute} = \text{Returns after diffusion of an innovation} - \text{Cost of making an innovation possible}$$

Formula 8: ROIn (absolute)

$$ROIn_{relative} = \frac{\text{Returns after diffusion of an innovation}}{\text{Cost of making an innovation possible}}$$

Formula 9: ROIn (relative)

Returning once again to a key point: Our aim with this book is not to make any claims in terms of "definitiveness and conclusiveness." This means that our formulae are also only intended to provide stimulus and ways to approach the term "InnovationQuality," even in arithmetic terms, and that on the basis of a short, to-the-point and coherent representation of the connections with suitable components, we can make a tangible contribution to the hopefully fruitful "discussion after reading."

We favor the indicator "Value creation through innovation (cVIn)." Return on innovation (ROIn) is, however, certainly also of interest as an indicator and there are many ways it can be used as an alternative or addition to "Value creation through innovation." However it should be noted that both indicators are based on a fundamentally different logic: ROIn is based on the logic that an innovation can be expensive or cheap. ROIn thus implicitly questions whether it is actually worth being innovative in the first place. The indicator "cVIn" is based on the logic that, in principle, an innovation is priceless, since, with such changes to something existing, sooner or later a company would not only fall behind but also perish. With the indicator "cVIn," the implicit question is with which new means and, temporarily, to what extent a company is safeguarding or expanding its future competitiveness.

A final point we wish to make quite clear: In an entrepreneurial and economic context, the "monetarization" of outcomes is the method of choice to identify the lowest common denominator for individual outcomes. Of course in other systems or contexts, a specific and generalized medium has to be used. So to evaluate an innovation in terms of overall society, the outcome for individual innovations could be looked at with the common denominator of "happiness," "contentment in life," and so on, assuming that the original aim of a society is actually for its members to be or become "happy" or "content in life."

2 The term value creation as a prospective hope[13]

In a prospective sense, to define value creation through an "cVIn" innovation, another indicator should be added, namely: the probability of something occurring – or "PIn" – that such value creation through innovation should happen in the first place. Expressed as a formula, this means that the indicator "cVIn" – both in an absolute and a relative sense – has to be multiplied by the indicator "PIn."

$$cVIn_{relative} = \left(\frac{\textit{Outcome after diffusion of an innovation}}{\textit{Outcome before diffusion of an innovation}} \right) \times \textit{Probability of occurrence of outcome after diffusion of an innovation}$$

Formula 10: Prospective determination of value creation through innovation expressed by "cVIn" (relative)

$$cVIn_{absolut} = \left(\textit{Outcome after diffusion of an innovation} - \textit{Outcome before diffusion of an innovation} \right) \times \textit{Probability of occurrence of outcome after diffusion of an innovation}$$

Formula 11: Prospective determination of value creation through innovation expressed by "cVIn" (absolute)

Of course the question is: How does one determine the probability of occurrence "PIn," which shows whether and to what extent such value creation will be achieved with an innovation? In the course of history, humanity has experienced many things, and in many different ways: radical political, social, cultural change, extremely disruptive invention and insights, as well as huge blows of fortune caused by nature and mankind. Without question, the world continues to change today, on a large or small scale, in a sweeping fashion but also primarily in a quick fashion and in many ways. There's no doubt that people today thus live in a time of change, in which many things – and in some

13 At this point we have to count on your curiosity and perseverance as we would like to derive and discuss further formulae, which we also consider important.

areas possibly even all things – are different from the things previous generations had to cope with. We are currently experiencing an epoch which is being heavily influenced by megashocks and megatrends. Something both of these phenomena have in common is the depth of effect (i.e., that change is substantial, existential and drastic) and the significant width of effect (i.e., that change extends to broad sections of a society or cultural groups, the whole of humanity, or even all life on Earth or the planet itself). Something that initially differentiates both of these phenomena from one another is the temporal dimension of change and their visibility in the here and now.

A megashock can be defined as a sudden, more or less unexpected and dramatic occurrence; a megashock is to a certain extent a major, startling jolt that provokes a sudden change. As a result of a megashock, entrenched opinions and irrevocable beliefs become completely unfounded or irrelevant from one moment to the next. This does not necessarily mean that it is some kind of disaster; even so-called "disruptive innovation," groundbreaking inventions or findings can initiate such shocks, for example an elixir that would mean an extension in human life, a vehicle drive capable of speeds in light years, the discovery or even contact with extraterrestrial life, etc.

A megatrend denotes something that is visible in the here and now, or at least the outline can be seen with a powerful, sweeping and constant effect; a megatrend is to a certain extent a continual erosion of the existing, with a complementary development of the New that provokes gradual change. As a result of megatrends, bit by bit, entrenched opinion and irrevocable beliefs more or less become completely without foundation or irrelevant.

Megashocks are rare occurrences. Although it appears that their probability of occurrence in terms of defined time periods such as centuries, decades or even years has actually increased in recent years. Naturally, we have to recognize that the chances of a sudden end to mankind cannot be completely ignored. And of course the same applies to the possibility of a sudden, worldwide stroke of luck, a possible megashock which we as humans typically tend not to consider. However, the probability of occurrence of such megashocks is and still remains small. We notice, especially today and especially in affluent societies, a major gap between, on the one hand, fatalistic reports which are presented by the media and depend on subjective perception, and, on the other, objectively justified threats. Stated more clearly: The frequency and intensity with which we think about apocalyptic scenarios is in many cases disproportionate to how often they actually occur, or at least to the scale on which they occur.

There is disagreement among experts throughout the world as to whether megashocks caused by nature have increased in recent times. Nonetheless, we are of the opinion that any intervention from mankind – especially as grave as they have been in recent centuries – naturally cannot be without consequence and that it is quite probable that, over and above "statistically normal" natural disasters, more may happen – provoked by the hand of mankind. As far as these societal and cultural megashocks are concerned, since time immemorial the world has been an unsettled place with wars or conflicts, technological or human failures culminating in catastrophic results, revolutions, counterrevolutions, changing paradigms in philosophical or technological terms. Human history was and still is a history of permanent and sometimes brutal change. Especially if one considers the last decade, one has the impression that even sections of society living in developed and comparatively stable countries – in political, social and economic terms – are undergoing a change with one crisis after the other: the collapse of the NEMAX50 on the German stock exchange, the subprime mortgage crisis, the sovereign debt crisis in Europe and the USA. We believe that one thing is indeed true: The more complex the systems that we use to make the world "our" world, the more likely megashocks become. One example from the currently much-scolded banking industry shows that even attempts to reduce the risk of a financial crisis could in themselves lead to new risks of other types of megashocks:[14]

> *Two little girls are playing on beach by the Baltic Sea. One of them shouts, "Bet I beat you into the water!" and rushes off. The other runs behind her and, lagging far behind, snorts: "I'm not playing anymore!"*
>
> *A clever reaction: Get out of it quickly before you lose. When banks do this they call it "risk management." [...] [But] this poses a fundamental question: Does introducing instruments to gauge risks and gain transparency actually make things safer? Or do they make the world more insecure? [...]*

14 The fact that risk minimization on one hand can result in new risks on the other was already made apparent in an outline given by the sociologist Ulrich Beck in his work "Risk Society: Towards a New Modernity " (1986).

Going back to the two girls on the beach: Risk management means getting out of it at the right time. If all the banks quickly identify the risks they face and maybe even use the same models to assess risk, then they will all more or less shout "I'm not playing anymore!" at the same time. And the market collapses. That's exactly what we saw in the last financial crisis. But if instead, risk were more non-transparent and the methods more imperfect, foundering markets would probably capsize later and also more slowly. Quite possibly, the ultimately effect would be that less damage is caused.

Which brings us to a realization: Probably all the progress that has been made to combat risk serves mainly to avoid a crisis that would be similar to the last one. But maybe it already sows the seeds for another, more complex crisis. (Translation, Wiebe 2012)

It is the very nature of a megashock that it is like a big bang. It has to be noted that it is not just because of the media, but also because it is the general nature of human beings that when such a big bang comes, we direct our entire attention to what has happened. As fate would have it, this results in us being so preoccupied that we lose track of other far-reaching developments, some already happening or looming large.

Of course knowing that megatrends happen and that megashocks are possible at any time has a serious influence on whether and to what extent entrepreneurial activities are successful. A megashock – e.g. a disruptive realization or invention – can erase all medium to long-term planning and expectations from one moment to the next. Even diversifying entrepreneurial activities will only make things appear safe, since ultimately this is based on the premise that known risks are being hedged against; there is no way, however, to hedge against the indeterminate, the unpredictable, the unknown. Given megashocks, the probability of occurrence of value creation through innovation or any kind of company project would have to be set at zero or almost zero. But naturally a company cannot and should not continuously be pulled down by such fatalism, since – and this needs stating clearly – in that case, one might as well completely "forget it."

A much more fruitful way to deal with the constant uncertainty of the future is to focus attention on megatrends, and, in doing so, also think laterally. Of course the distinction between a megashock and a megatrend is sometimes on a purely analytical level: If one has a mechanistic picture of the world, even the most sudden occurrence does not appear out of thin air, but instead as a consequence of other events. To many who are more closely involved in key issues, even shocks resulting from new insights and inventions are certainly still astounding, even if they are not entirely surprising. A decisive factor with a megashock is that a large number of people are completely taken by surprise by the occurrence. Given this, things like the economic crisis of the last decade could be called a megashock. With hindsight, it is clear to everyone that the trend toward increasingly reckless financial transactions and permanently increasing sovereign debt inevitably and unavoidably had to result in a big bang. But at the beginning, the attention of many people was of course not focused on this trend, so the surprise, or rather the shock, was correspondingly significant. In many cases, megashocks are actually only a megashock for those who did not – some time ago – already recognize the development as a megashock and/or did not previously think laterally. In this regard, one could point to instances such as when the entire music industry, and with it the entire "content industry" (films, journalism, etc.), took a long time to notice the impact that digitization and networking megatrends could have on the economy in general, or more specifically on their branch of the economy. It just took a small number of lateral thinkers to find an ingenious way to link end devices with replacement applications and services. And before people knew it, for a long time the majority of the "content industry" was practically in a state of paralysis – or megashocked. In the meantime, everyone understands how, with inkjet printers, hardware producers can earn more through the ink than through the printers, so we wait with anticipation to see how shocked the majority of people will be about the possible consequences of 3D printers and look forward to the innovations of the (small number of) lateral thinkers.

For businesses, this also means that customer understanding is not confused with possessing plenty of data and facts about customers. Having a close understanding of customers as a basis for knowing if a customer might see the New as "good" is derived from having an understanding of the deep-seated need behind something new:

> *As a term, customer understanding should be relatively accessible and immediately understandable. If customer value is the ultimate goal of business activities, then one should know what it is based on and on what basis it can be delivered. Does this then mean that one should ask one's customers what*

a product should be like? Not necessarily – one would run the risk of just finding things out that stem from a re-composition of known product elements. Customer understanding should always bear in mind, or at least anticipate, which new (or even known) needs could become relevant for the customer in the future. Again, this also entails extrapolation. Here too, we encounter tasks that involve hypothesis and call for differentiation. Working exclusively with focus groups and customer surveys is a guaranteed way to achieve substitutable mediocrity. Steve Jobs is quoted as saying that he always objected to focus groups and consistently avoided them; despite this (or given what was said earlier, actually because of this) no-one would doubt that Apple products excel in their exemplary reflection of customer understanding. (Translation, Freund 2013: 21–22)

There is simply no sense asking customers about something radically new, i.e., about something that they know absolutely nothing about. Despite all the creativity and progressive thinking, even focus groups, early adopters and lead users are still more or less held back by incremental thinking and make extrapolations of the status quo.[15] A customer is quite conceivably the least appropriate person to speak to when the issue is radical innovation – this also (or particularly) applies to the quality of things that are hitherto completely new, because how should a customer be able to say if something is "good" if they have never seen it before, they have no frame of reference and they simply cannot interpret it.

A key success factor in understanding the probability of an undertaking – and thus also an innovation – becoming a success is to constantly keep one's eyes and ears open, thinking further and "more laterally" than the competition, and thinking further than the customer. The future nature of the world – and future requirements, or the nature of requirements – needs to be analyzed and anticipated. The art is to not just concentrate on the existing trends that are already understood, but also to notice trends that are just beginning to emerge – or even set the trends. Doing this requires intensive thought about the future ("future-oriented research"). The hypotheses and models that this results in still have to be systematically linked to objectives and strategies of the company ("future-oriented management"). And finally, a foundation is required, based on a culture and climate that not only allows people to think laterally, or make mistakes and "fool around," but actually encoura-

15 Another practically insurmountable hurdle aside from this incremental thinking is that these customers groups have an inclination for solutions that tend to be utopian or, for example with product innovations, are at least not implementable in technical terms in the long term.

ges this. In organizations and societies where people (have to) stick to "the rules," the New does not develop.

Thus, as a result, for the value of "probability of occurrence of value creation through innovation 'PIn'," what we are dealing with is a highly complicated and complex factor that is shaped by opportunities and risks, expectations and indeterminacy.

The probability of occurrence "PIn" is thus an indicator somewhere between 0 and 1, concealing an intense and elaborate occupation with the future, just as much as a notion that anything can be possible – as well as the opposite.

It is thus self-evident that we can not provide a formula for this factor at this point; there is no sensible way to predict the future with any certainty. Not even some kind of "theory of everything" could do this, since firstly it would consist of virtually endless numbers of variables and secondly any such formula would never be complete – in fact many variables could just pop up at the drop of a hat and change their magnitude. Such formulae may allow us to state the future probability of the occurrence of risks and opportunities that we know, but these formulae quite simply say nothing about the indeterminate and the unknown that await us in the future. And if nevertheless we did know of such a formula, then we would certainly not reveal it in such simple terms in the subchapter of a book...

3 The timeframe of value creation

Also already mentioned, and, in our opinion, quite rightly stated in a "penetrating" fashion, Schumpeter defines "innovation" as: "the doing of new things or the doing of things that are already done, in a new way" (Schum- peter 1947: 151). Central to this definition is the use of the verb "doing": It is not just about having an idea or notion of the future in one's thoughts; it is about translating these ideas into actions and allowing new knowledge, or knowledge not applied in this way before, to become a value-creating reality. An innovation thus goes much further than an actual idea or invention, as it entails additional activities that dictate market success. An innovation in Schumpeter's sense is thus also more that the simple implementation of an idea, it is more than a prototype, product study, etc. The implementation of an idea does not yet create value, nor reduce costs, nor access new customer segments, etc. Something only becomes an innovation when an idea is not just implemented but also successfully introduced, made available to society and it is made suitable to requirements, etc.

Morten T. Hansen and Julian Birkinshaw (2007) make differentiations in the innovation process with their model, the "innovation value chain," which includes three sub-processes: idea generation, conversion and diffusion . As already explained in the introduction of this book, aside from acquisitions or M&A activities, innovations are the strategic means by which the future viability of a company is safeguarded, extended, or both. The term "innovation value chain" also implies that an innovation initiates and/or enhances and/or sustains value creation. And to state this once again: For companies, the function of innovations in the first place is that either money comes in, or more of it comes in, or it continues to come in. Little wonder therefore – and this is not just valid for the model of Hansen and Birkinshaw but also for many other well-known models (e.g., Rogers 2003) – that the innovation process ends with diffusion. Yet for a company, the innovation process actually does not end until the precise moment the customer or recipient uses the innovation and a feedback effect results for the company or originator (e.g., payment flows with product innovations, cost savings or efficiency gains with organizational innovations, etc.). As a result, the innovation process should at least really be extended to include this feedback effect.

The originator of an innovation hopes that the diffusion of an innovation will result in some kind of feedback effect; the starting point for this effect comes when the recipient uses the innovation. But why does the recipient use an innovation? A somewhat tautological answer to this would be: Because it is of use to him, because the innovation promises to create some kind of value for the recipient, because the innovation results in something of value being established, extended or maintained for the recipient. Value creation for the recipient is not mentioned in the literature, but always pointed to implicitly when it is mentioned that customer value should also be a priority when it comes to any kind of entrepreneurial undertaking. Indeed, no company in the world can exist in the long term if it fails to focus its business operations and strategies on customer value.

> *Peter Drucker (subsequently also continued by Fredmund Malik) described the purpose of entrepreneurial activities simply, yet extremely accurately as: "the creation of customer value." If, compared to the competition, a company satisfies a new customer need or an existing one in a superior way, either the market will expand as a whole or market share will shift to the advantage of the "better" provider. Discussion often revolves around Drucker's definition, since market success, shareholder value, returns (or in a more general sense, the achieved profit) are not seen as the ultimate purpose of companies. Instead, for Drucker, economic parameters constitute fundamental prerequisites without which a company cannot exist in the long term. As such, they perform more of a diagnostic function. (Translation, Freund 2013: 5)*

Why should customers in a market economy want to use products and services that provide them with no value, that result in no personal value creation? The objective of "market success" is only achievable through the means of "customer value." Value creation for the recipient of an innovation is thus not a completely different way of looking at things, but a complementary one with respect to value creation for the originator of an innovation. Coming back again to an important point: A newly developed product is only an innovation when it results in the generation of turnover; turnover is only generated when a customer buys this new product; the new product is only bought by the customer because he hopes that the new product will satisfy a need, a preference or desire. It can only be said that something is an innovation if something new or something changed brings about value creation for the recipient and, in return, he brings about value creation for the originator of the innovation through feedback effects. A complete innovation value chain would therefore look like this.

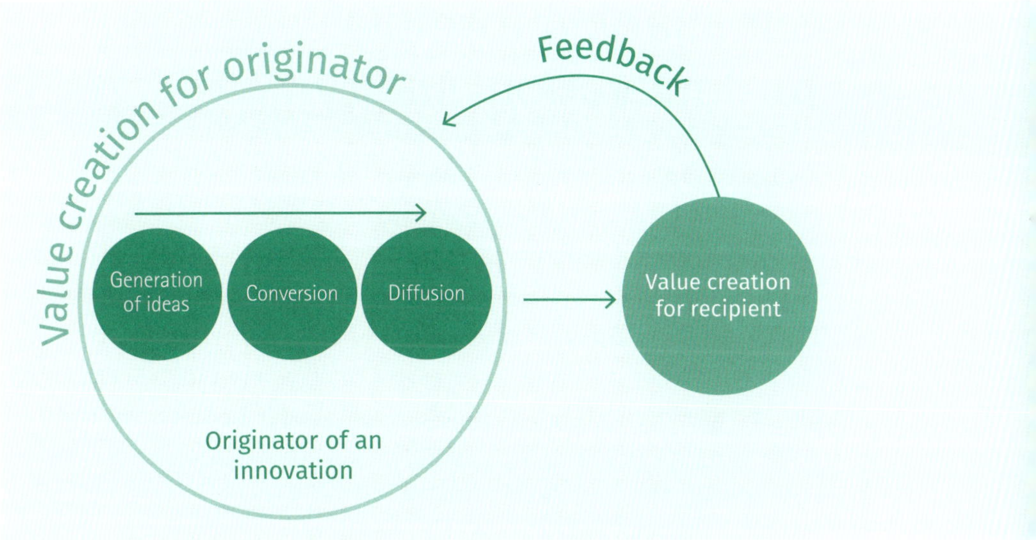

Figure 22: The complete innovation value chain

Thus, the actual value creation resulting from an innovation does not begin with diffusion, but instead only at the point when positive feedback happens on behalf of the recipient of the innovation. Again, stating this clearly: An indication of value creation through innovation is not that it has been diffused, but instead that the money is coming in.

A temporal indicator that reflects when there is value creation through innovation is thus actually not the precisely definable point of diffusion, but instead the abstract and innovation-specific timeframe ("t") of feedback.

This indicator is abstract because a company must define for itself when, or by when, such feedback effects should take effect. For example, one could set a deadline from the moment of diffusion, that within x months or x years this or that value creation should be achieved. The indicator is innovation-specific because to a certain extent different types and kinds of innovation also have different "rates of maturity": So value creation that is possible through product innovation tends to be perceived more quickly than value creation that is possible through organizational innovation.

4 Summary: A formula for InnovationQuality

Expressed in general terms, the outcome of an innovation is that after its diffusion, i.e., after sharing, there is a change – or delta "Δ" – compared to the outcome before diffusion of the innovation. This relationship can be expressed in relative or absolute terms, i.e., either as a quotient or as a difference of this change. Furthermore, change resulting from an innovation can be seen in a retrospective or in a prospective sense. Overall, as a result, "InnovationQuality" can be expressed (in a retrospective sense) with the following formula:

InnovationQuality "InQ" is the value creation delta resulting from an innovation within the timeframe "t".

$$InQ_t = \Delta cVIn_t$$

Formula 12: InnovationQuality "InQ" (retrospective)

Looked at from a prospective angle, naturally there is uncertainty and indeterminacy to be dealt with. As a result, when the focus is on the future, a probability factor should be included. Overall, as a result, "InnovationQuality" can be expressed (in a prospective sense) by the following formula:

InnovationQuality "InQ" is the value creation delta resulting from an innovation within the timeframe t multiplied by the probability of occurrence "PIn" for this delta within this timeframe t.

$$InQ_{t\text{-prospektive}} = \Delta cVIn_t \times PIn_t$$

Formula 13: InnovationQuality "InQ" (prospective)

Often the organic growth of a company is not just down to an individual innovation but it is the sum of several innovations.

Thus in a similar fashion, Corporate InnovationQuality ("Corporate-InQ") is the sum of all value-creating changes resulting from innovation.

Care should be taken to ensure that, whenever possible, all innovation activities within the company are considered, i.e., all kinds of innovation.

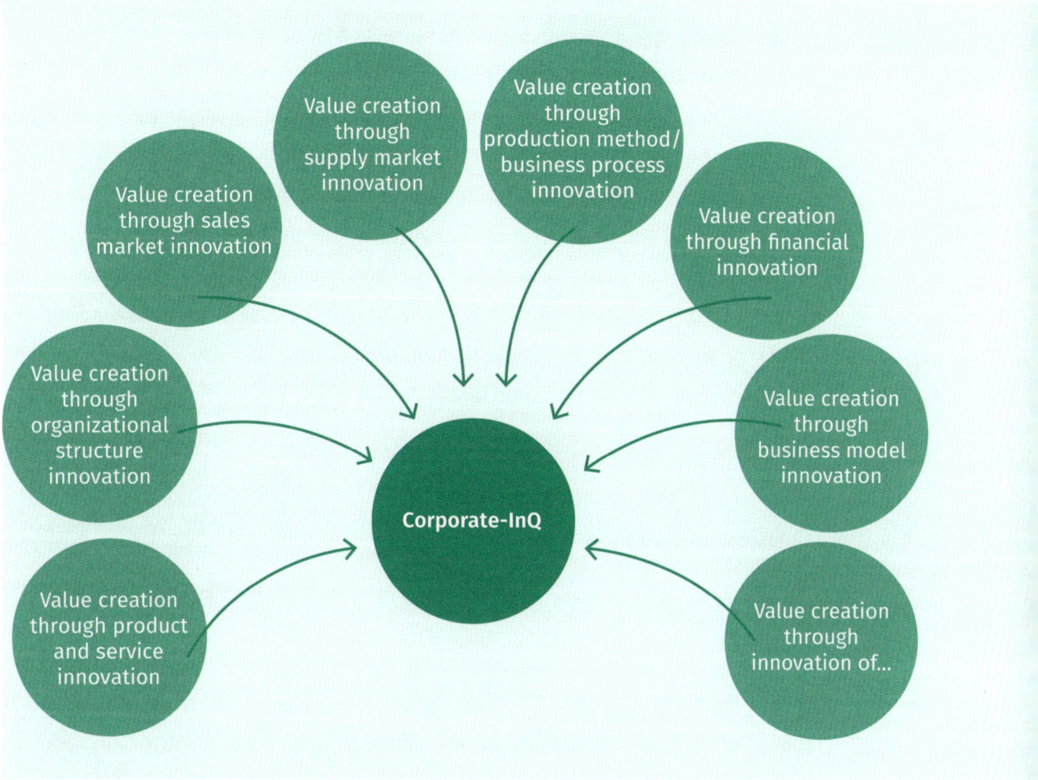

Figure 23: Corporate InnovationQuality "Corporate-InQ"

The degree of Corporate-InnovationQuality can in turn be expressed retrospectively as a key performance indicator or prospectively as a target figure for business objectives. With the retrospective calculation of corporate InnovationQuality "Corporate-InQretrospective", value-creating changes resulting from innovation – 1, 2, X… "ΔcVIn1, ΔcVIn2, ΔcVInX…" – are added to one another within the timeframe t. Naturally, to do this, the achieved outcome of each instance of value creation has to be converted into the same "currency," so: Within an entrepreneurial context, each specific outcome resulting from the diffusion of an innovation has to be monetized, i.e., expressed through the general medium used in the economy: monetary value. Furthermore, the subsequently monetized outcomes must be the same type of outcome, i.e., outcomes can naturally only be added to one another if they all relate to the same kind of business outcome (turnover, profit, revenue, etc.).

Of course, aside from this economic effect it can also be suggested that there are other desired outcomes, such as psychological effects (customer satisfaction) or behavioral effects (e.g., customer loyalty). We have decided to use the economic effect as this not only stands at the end of the causal chain but is also at the top of the pyramid of objectives, i.e., the psychological and behavioral effects preclude the economic outcome in causal terms as a necessary precondition. Despite this, all of these effects are ultimately useless for a company if there is no economically relevant action (purchase, contract, etc.) at the end. Expressed another way, the aforementioned psychological and behavioral effects may be pertinent to the determination of InnovationQuality, but these are not as meaningful as the economic effect. As a result, we adopt the following formula for corporate InnovationQuality:

$$Corporate\text{-}InQ_{t\text{-}retrospective} = \sum_{i=1}^{n} \Delta cVIni_t$$

Formula 14: Corporate InnovationQuality (retrospective)

For the prospective objective planning of Corporate-InnovationQuality (Corporate-$InQ_{t\text{-}retrospective}$) the value-creating changes that result from innovation 1, 2, X... "ΔcVIn1, ΔcVIn2, ΔcVInX..." are added to one another within the timeframe t, and then each is also multiplied by the corresponding probability factor for the probability of occurrence of value creation "PIn" (which resulted from innovation in this timeframe t).

$$Corporate\text{-}InQ_{t\text{-}prospective} = \sum_{i=1}^{n} \Delta cVIni_t \times PIni_t$$

Formula 15: Corporate InnovationQuality (prospective)

If you are still with us at this point, thanks for your perseverance!

Some concluding comments, should we be accused of being over-simplistic 5

Ours aims with this process of derivation – and the formulae we have derived – are to define InnovationQuality as a general, overall pragmatic term. With such an undertaking, what has been required until now is a simplification bordering on the naïve. For example, with the retrospective definition of InnovationQuality, it is of course all too simple and naïve to assume that changes that occurred after the diffusion of an innovation were also causally linked to the innovation's diffusion. Though it is possible that the changes that occurred after the diffusion of an innovation (e.g., in turnover or profit) were caused by something completely different, and that they simply happened to coincide with the innovation. Every company that encounters success or failure should ask: What were the reasons? What were the factors or principles for the success or failure? The fact that any such reflection can never be final, that one can thus never be ultimately certain about the actual reasons, may be an epistemological dilemma but it is not an excuse to nonetheless try. Furthermore, not only companies but also individuals should by necessity accustom themselves to the notion that an outcome stems from one's own actions. If we entirely ignore this idea and consider all outcomes to be a coincidence – the result of "forces unknown" – we lose any certainty regarding our personal effectiveness and thus any certainty that our actions can make a difference, for example between success and failure.

With the prospective definition of InnovationQuality, we perhaps still have to be bound by simplicity and naivety and reconcile the completely unknown (future value creation through innovation) with another completely unknown (the probability of this value creation actually occurring). Naturally, the result of multiplying values with one another whose nature one can only hope to guess at, can only be aspirational or a general estimate. This is not on the same scale, but all stockholding companies throughout the world will be familiar with what it feels like issuing some sort of forecast for possible annual results; the same applies to all economic forecasts of GDP growth. The reason why, despite all academic and logical absurdity, we still choose to believe in prospective determination of InnovationQuality is this: He who sets no innovation objectives, will not lag behind the competition, but succumb to it.

The prospective determination of InnovationQuality is therefore less about precisely estimating the future value creation that results from an innovation. Instead, with prospective determination, it is more about being clear in the first place that future value creation is also, or above all, a result of innovation. As such, the indicators we provide do not offer scientific precision or objectivity. For a number of decades, the science of business – and to a certain extent also management – has been marked by a kind of positivism, a "mono-cultural" belief in facts to the great detriment of instinctive behavior, imagination and creative thinking – and with this, to the detriment of many things that are precisely what entrepreneurship should be all about. In case we are misunderstood: This is not "revenge" for the introduction of scientific thinking in business, and this is not an attempt to drive such thinking out of business! But the aim with the formulae for InnovationQuality is to express this dichotomy: On the one hand there are a growing number of sciences that are dictated by empirical thinking and who, in their very essence, only have access to numbers from the past and present, which, not wanting to turn one's back on empirical methods, can only be extrapolated into the future, and which can of course then be continued ad absurdum based on any undefined and unpredictable outcome. On the other, there are companies that are increasingly influenced by permanent and sometimes powerful change, whose future survival sometimes lies precisely in their ability to make a radical break with their own past or present, without having any reliable numbers to turn to and thus not knowing if their courage will even be rewarded.

Step 1: Introduction to the two terms of “Quality” and “Innovation” by closely examining the phenomena and their distinctive characteristics

Step 2: Combination of the two terms, or the two phenomena of “innovation” and “quality”

Step 3: Specific definition of the interpretation of “ InnovationQuality”, also as a quantitative variable

Step 4: Outline of a model for InnovationQuality

Step 5: Illustration of an approach for managing InnovationQuality

Using our term of InnovationQuality as a starting point, we outline a model for InnovationQuality in the following. Although we have created this model primarily from the standpoint of businesses, the aim of the model is to be compatible with non-business contexts.

A look at some of the existing innovation models 1

Of course there are already a multitude of models that attempt to capture the phenomenon of "Innovation" or even make it tangible. The innovation models that appear to be pertinent in this context are the ones that are used in (inter) national reports on the innovative activities and capability of countries and regions, for example, the Innovation Union Scoreboard of the European Union or the annual report of the Expert Commission on Research and Innovation (EFI) issued by the German Federal Government, as well as a variety of other national reports. Also, there are innovation models that are particularly pertinent because they are used as a basis for international rankings of the competitiveness and innovative capacity of domestic economies. These include the Global Competitiveness Report of the WEF, the Global Innovation Index of INSEAD Business School and Innovationsindikator Deutschland, which is issued by the Deutsche Telekom Foundation and the Federation of German Industries (BDI). These reports and rankings may differ in terms of structure, but one thing they all have in common is that they are based on a more or less elaborate and conscious innovation model. This theoretical foundation is used to derive indicators. Each individual indicator provides an insight into the nature of relevant components of innovative activities and innovative capability. And on the whole, these indicators provide an insight into the overall nature of innovative activities and innovative capability. The particular relevance of innovation models and the indicators that are derived from these models lies in the following: Not only the models but also the indicators are not merely instruments with which to add structure to excerpts of a reality, thus making it observable; the models and the indicators in themselves serve to give structure and perform a function: Models and indicators make it possible to determine relevance, focus attention and ascertain meaning.

If an indicator is considered a particularly important yardstick for innovative activity and innovative capability, this naturally creates awareness that by influencing this, a significant change can be made to one's own innovative activity and innovative capability. So, for example, if the indicators "expenditure in research and development" or "number of patents" are considered particularly relevant, this creates the impression that one should work more on this indicator – in other words, one could raise the amount of money spent on R&D and the number of patents to change one's innovative activity and innovative capability.

A major problem with innovation models and the indicators that are derived from them is the dilemma of empirical research, namely the conflict between "rigor and relevance": Things that are easy to conduct empirical research on are often not relevant and things that would be relevant are not easy to research empirically. So supposedly, the selection of indicators for measuring innovative activity and innovative capability will be strongly influenced by the fact that there is statistically useable material. Taking this to the extreme: It is not the sense and meaningfulness of indicators that dictates whether they are accepted, but their existence, or easy access to available hard data (i.e., data that was gathered quantitatively). Thus, in the reports and rankings named above, one indicator that is frequently used for innovative activity and innovative capability in an economy is the number of registered patents. Patenting activity can be captured relatively easily and objectively, but the objectivity and ease with which an indicator can be measured says nothing about how meaningful it actually is. We are not questioning the requirement to protect intellectual property rights, but it should be noted that concentrating on patents creates the following impression: (1) It appears that there is a causal relationship between the number of patents and the number of innovations; (2) It appears that the innovation process is about registering patents; (3) It appears that the rule is, the more patents there are, the more innovative people are. First and foremost, these days, patents are not being used to establish the necessary safeguards on intellectual property; in many cases they are a protective shield and a passive blocking device or an active weapon against competitors. We also agree with Schumpeter when he writes:

> *[...] The inventor produces ideas, the entrepreneur "gets things done." [...] An idea or scientific principle is not, by itself, of any importance for economic practice. (Schumpeter 1947: 149)*

The new combinations are always present, abundantly accumulated by all sorts of people. Often, they are also generally known and being discussed by scientific or literary writers. In other cases, there is nothing to discover about them, because they are quite obvious. [...] It is this "doing the things," without which possibilities are dead, of which leaders' function consists. [...] Economic leadership in particular must hence be distinguished from "invention." As long as they are not carried into practice, inventions are economically irrelevant. And to carry any improvement into effect is a task entirely different from the invention of it, and a task, moreover, requiring entirely different kinds of aptitudes. [...] It is, therefore, not advisable, and it may be downright misleading, to stress the element of inventions as many writers do. (Schumpeter 1934: 88-89)

This opinion is underscored empirically by a broad study carried out by the management consultants Booz Allen Hamilton: It was not possible to ascertain a correlation between expenditure on R&D, patents and performance indicators such as turnover growth, profit growth, profitability itself and market capitalization. Ultimately, this means: "[...] There is no correlation between the number of patents and financial performance" (Scanlon, 2006). In many cases, a patent is nothing more than an idea that has been given the required legal protection; but such an idea is primarily nothing more than the possibility to create (but also block) an innovation and with this, the possibility to realize financial growth. A patent lays the foundation for an innovation if it is exploited in such a way that it not only brings about success, but also – in a narrow (radical) sense – it brings about success that upsets the equilibrium.

Another criterion that supposedly has a decisive influence on whether an indicator appears in a report or ranking is whether factors that underlie the indicator can be positively influenced. Taking this to the extreme, indicators not only have to provide a clear representation of the processes surrounding the phenomenon of "innovation," they have to be usable within a political or business context. (cf. Weingart 2011: 21) So it is hardly surprising that the indicator "expenditure on research and development" is a regular on the list of all rankings and reports. The underlying thought behind this is that innovations primarily – and sometimes exclusively – originate in this area and that there is a causal relationship along the lines of "lots of investment = lots of innovation." With the arguments in the above-named rankings and reports relating to the use of implicit or explicit innovation definitions, it should be noted that strong emphasis is given to technical innovation – even if some reports and rankings maintain the opposite. The indicators that are intended to be used to measure the innovative capability and innovative activity of an economy clearly lay the focus on this type of innovation (Mergenthaler, Faix 2014). Also, it should be considered that, in itself, the level of expenditure on "research and development" says little about the innovativeness of an economy or a company.

> *A central indicator for steering the innovation policies of the EU and the OECD member states is the correlation between R&D expenditure and national output (GERD/GDP), which is also a guideline set by the EU in the Lisbon Strategy (3 percent). This may provide an indication of the willingness of the affected governments to fund research, but this indicator says practically nothing about the actual suitability of funding policy to achieve innovation targets. The differences between the aggregated expenditure of industry on R&D (BERD = Business Expenditures for R&D) also do nothing to explain the innovativeness of a domestic economy, but instead are a reflection of the differing degrees of research intensity within the industrial infrastructure of different countries […]. The different R&D levels of the economy in different countries are also a reflection not primarily of the degrees of research intensity, but mainly of the development intensity within the industrial infrastructure of countries. The "D" part of -expenditures accounts for the lion's share of R&D spending. In Germany, around a half of R&D expenditure in the economy is in the automotive and supply industry, and this is mainly accounted for by inexpensive material and crash testing. (Translation, Weingart 2011: 20)*

It would be safe to assume that conventional innovation models were developed in the following way: (1) One given is a more or less elaborate theory regarding the situation surrounding the phenomenon of "innovation" – the structures and processes. (2) This situation is subsequently examined to see if it can be observed in a quantitative and empirical way, and if it can be influenced on a political or entrepreneurial level. (3) Aspects that fulfill the conditions under (2) are then connected and synthesized into an innovation model. Naturally, such an approach is entirely correct from a scientific and theoretical standpoint, but one does have to ask whether it is pertinent beyond necessary political influence. There is an issue here on two levels relating to rigor and relevance, as mentioned above. (1) Should a criterion relating to whether a situation can be properly observed using empirical and, more importantly, quantitative techniques really dictate whether – and to what extent – this circumstance provides a foundation for establishing an innovation model? (2) To what extent do the aspects that were pre-selected in this way actually have an influence on the innovative activity and innovative capability of an organization or a community?

Models – and aspects that relate to a situation or are derived from these models – fulfill a role in (corporate) policy and management by making it possible to control the system. This control is based on the following train of thought: The existence and essence of certain aspects reflect the fact that a desirable outcome will become more likely. The existence and essence of such aspects are a reflection of target variables that, indirectly, should be achieved: If this or that aspect is evident within processes and structures, to this or that extent, then it is probable, to this or that extent, that a desirable final outcome will be achieved. So the more precisely a certain aspect can be captured, the more precisely one can determine the probability of a successful final outcome. This is the basis of something like the "logic" that more patents will enhance the likelihood of innovations, despite the fact that innovation follows its own logic, which is completely different and unpredictable:

> Enterprise A owns a variety of patents, albeit all of an incremental nature, whereas Enterprise B only holds one patent which has the possibility to be radical or even disruptive.

> At Enterprise A, a large number of people are working in the research and development department on a large number of innovation projects, whereas there are only two people at Enterprise B working on a single project in a converted garage, but this has the possibility to be radical or even disruptive.

At this point, it is worth reminding ourselves that businesses and domestic economies are highly complex and highly dynamic systems and not machines that follow banal, mechanistic laws of causality. If a system is complex, by implication its components, constituents, etc., can no longer be viewed in isolation, instead they have to be considered in their transactive, networked totality. Totalities are different from purely additive or purely hierarchically merged entities in that the contributions made by their constituents are not summative or more refined hierarchically. Instead, the result is "over-summative" – culminating in a transactive network whereby the whole is more than the sum of all parts. The fact that a system is dynamic means that it can develop in a completely unpredictable, if not chaotic, way due to the sensitive interdependence of initial values. What this means is that even the most minute differences in the original state of a business or domestic economy will result in so-called deterministic chaos, producing completely different behavior within the system over time.

The chances of innovation and the possibility of unearthing true gems of novelty are of course raised by registering more patents and investing more in research and development, but it may also just take a single, ingenious insight. Naturally, quantitative variables also make a powerful statement, even when it comes to innovation, but this statement should not be reinterpreted as a crude cause-and-effect relationship, such that a simple quantitative increase is also tantamount to an innovative outcome. Unexpected novelties, eureka moments, the game-changing idea. These are all the sorts of events that most certainly can just confuse people in one way or another if variables are introduced for quantitative reasons. But if they are, then this should not be in a crude or mechanistic way, as in "more of this results on more of that." Innovation – and being clear about this is important to us – is an unruly, capricious and complex phenomenon that cannot be tamed with simple if-then formulae.

The consequences of all the lessons we can learn from this are:

> With innovations, quantitative targets and control variables can at best reflect a correlation or indicate a tendency that more or less of a particular indicator would result in the possibility of an innovation, or enhanced InnovationQuality.

> Primarily, a model for InnovationQuality should give preference to qualitative information. Naturally, such information is of no more use in "controlling" innovation, but with such qualitative variables innovations can be suitably captured in terms of their unruliness, capriciousness, complexity and dynamism. (Bormann 2012)

> A model for InnovationQuality is contradictory to a somewhat naïve "technicism" – i.e., the concept that one simply has to transfer the methods of successful innovations to one's own innovations. So general success factors cannot be read into qualitative data – at the very best, only design principles can. Instead, it must be clearly understood what results in an innovation, why, how and within which context in terms of

timing and the social context. Thus, a model for InnovationQuality can and should basically only include categories of indicators but not specific indicators. And a model for InnovationsQuality cannot and should not say anything about which of these categories have which impact.

> A model for InnovationQuality is basically work in progress. In other words, it can and will change on a number of fronts in the future and possibly completely change its form. The form we have chosen and the criteria etc. that we have selected are thus not the result of dogma and make no claims in terms of "definitiveness and conclusiveness."

2 Derivation of a model of InnovationQuality from the EFQM excellence model

We derive a model for InnovationQuality in the following, drawing on the EFQM Excellence Model. One big advantage with the EFQM model is that, as a business model, it makes it possible to gain a holistic view of all kinds of organizations. It is thus not restricted to use with organizations in the manufacturing sector, nor does it provide a restricted view of the tasks and results of research and development departments. Another advantage with the EFQM model is that although it is another vertical tool of differentiation – i.e., it allows for specific indicator systems and indicators themselves – in its general form, it is still not restricted to certain criteria . On the one hand, this steers observation of the phenomenon of "innovation" in a number of general directions, creating initial awareness of the areas that actually need observing. On the other, the extremely general criteria used by the EFQM model mean that observations are not strictly channeled in one direction, so they always remain more or less open to the New and different. Finally, and we openly admit this, another advantage with the EFQM model is that it is an established and very well-known model; thus a model for InnovationQuality does not completely run perpendicular to the already known – instead, it is an "incremental innovation" so for some it may appear easier to digest to some extent.

The EFQM model covers two main criteria – enablers and results. These in turn are broken down into nine main criteria.

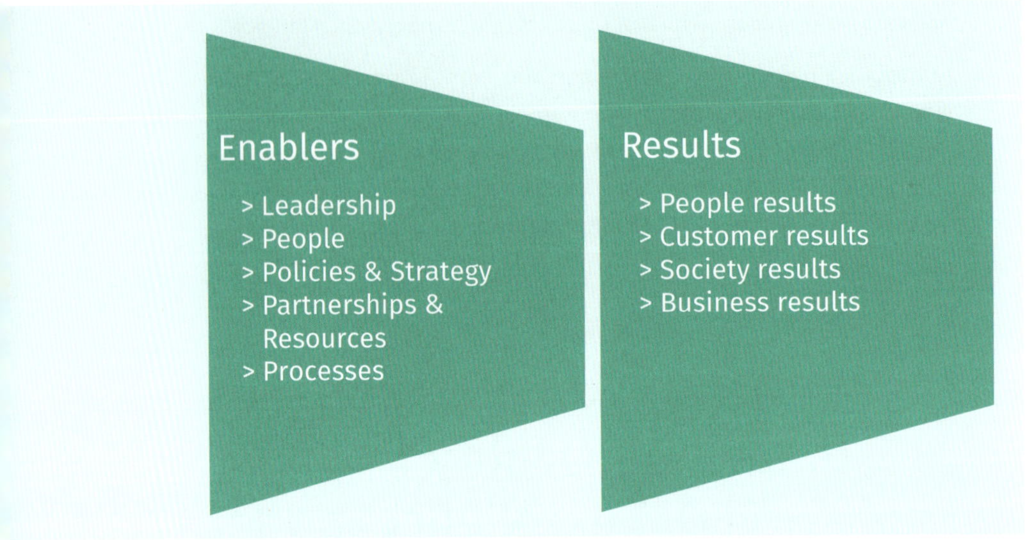

Figure 24: The main criteria of the EFQM Excellence Model

2.1 Enablers of InnovationQuality

It is immediately apparent that the criteria that come under the "Enablers" in the EFQM model almost exclusively relate to people and their actions. Schumpeter already emphasized that things that are new, different or better do not simply come down like "manna from heaven." Instead, it is the consequence of ideas that are translated systematically by people into a reality, i.e., initiated, planned, implemented and overseen. Without human action, an idea remains a flight of fancy. It therefore requires innovators – people who think things up, take the lead and make things happen. For people to be able and willing to innovate at a company, a company culture that fosters innovation needs to be established. For a company culture to foster innovation, firstly it takes a willingness within the company to try out and integrate the New, i.e., defining and working on innovative objectives should be part of the very essence of the organization. Secondly, obstacles that thwart innovation must be overcome and partnerships that promote knowledge sharing and collaboration inside and outside the company must be established. Thirdly, a company culture that fosters innovation should not just tolerate but actually encourage lateral and different thinking, and encourage people to explore new avenues: "From opposition arises progress." (Johann Löhn). And finally, managers have to move on from traditional leadership styles: To do this, for example, people have to be encouraged and challenged to organize themselves, creativity has to become a core constituent of the business and less formal, non-hierarchical, communication channels must be used.

All of these aspects illustrate factors which – drawing on the EFQM model – are central to the accomplishment of innovation and thus how InnovationQuality is shaped.

1. People who are capable of innovation

2. Managers who enable innovation

3. Processes and structures that are focused on innovation

4. Partners and resources that foster innovation

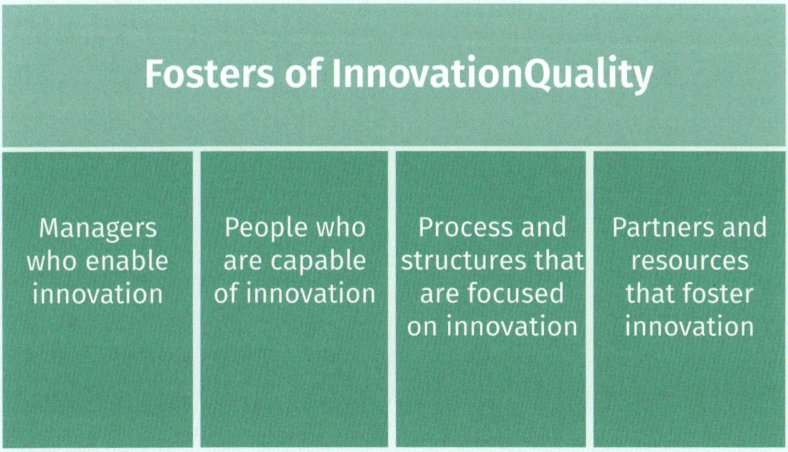

Figure 25: Fosters of InnovationQuality

With all of these factors, the existential difficulty in management lies in the antinomy, the fundamental contradiction of the poles of order and freedom. On the one hand, no organization can function without order, planning, predictability, central control, bookkeeping, instructions from superiors, obedience and discipline. An organization without order would not be an organization but an incoherent "rabble." On the other hand, innovation engenders a certain degree of disorder in the form of recklessness regarding rules and bureaucracy. This is because it is not without such freedom that the creative powers of imagination can actually unfurl. Managing innovation entails demanding as much existing order as necessary so one can simultaneously "shake-off the world of yesterday, in order to have the freedom to create the world of tomorrow ." (Peter Drucker, quoted by Haas Edersheim 2013: 105)

2.2 Reasons for InnovationQuality

The attentive reader may have noticed that the above list of the fosters of InnovationQuality omits aspects relating to "strategy/policy." Indeed, this aspect should only feature now – but only with a modified designation: Instead of "strategy/policy" the better term is "objectives," or better still, "innovation objectives." Entrepreneurial objectives play a pivotal role in corporate development (Faix et al., 1996). Objectives reflect the decision-making criteria of a company. In other words, the selection of possible actions is only possible by referring to the previously formulated objectives of the enterprise. Furthermore, company objectives provide helpful goal posts when it comes to orientation, coordination and the legitimization of company decision-making. Objectives and the desire to achieve objectives are the essential driving force of entrepreneurship and, with this, also a company. As a clean distinction is often not made in discussion between the terms "objectives" and "strategy," it is made clear once again now: The objectives of a company dictate strategic direction, taking into consideration potential and the overall situation of the company (framework conditions). To achieve a specific company objective, a host of individual actions must be carried out, all working in harmony with one another, having been selected from a variety of alternatives. A package of measures that was planned in this way and is aimed at realizing one or several objectives – i.e., the way to the goal – is a strategy. Company objectives represent guidelines that should be fulfilled through strategies as ways to achieve objectives. Because there are generally many different ways to achieve objectives, and because, as a result, there is a somewhat complex choice of individual measures, the company strategy is frequently afforded less attention than the objectives themselves in discussion, documentation, etc. This is simply a result of the nature of business – but despite this, even the best strategy is useless without objectives (cf. Faix 2008). An objective that is pursued with every possible intention is the reason why, afterwards, all necessary avenues are identified and all means are used.

The guidelines provided by an "innovation objective" stand at the beginning of the innovation process and they are the reason for every action relating to InnovationQuality. Subsequently stemming from this comes every effort to identify means and ways to develop and expand in order to achieve the objective. The outcome of all these endeavors is value creation.

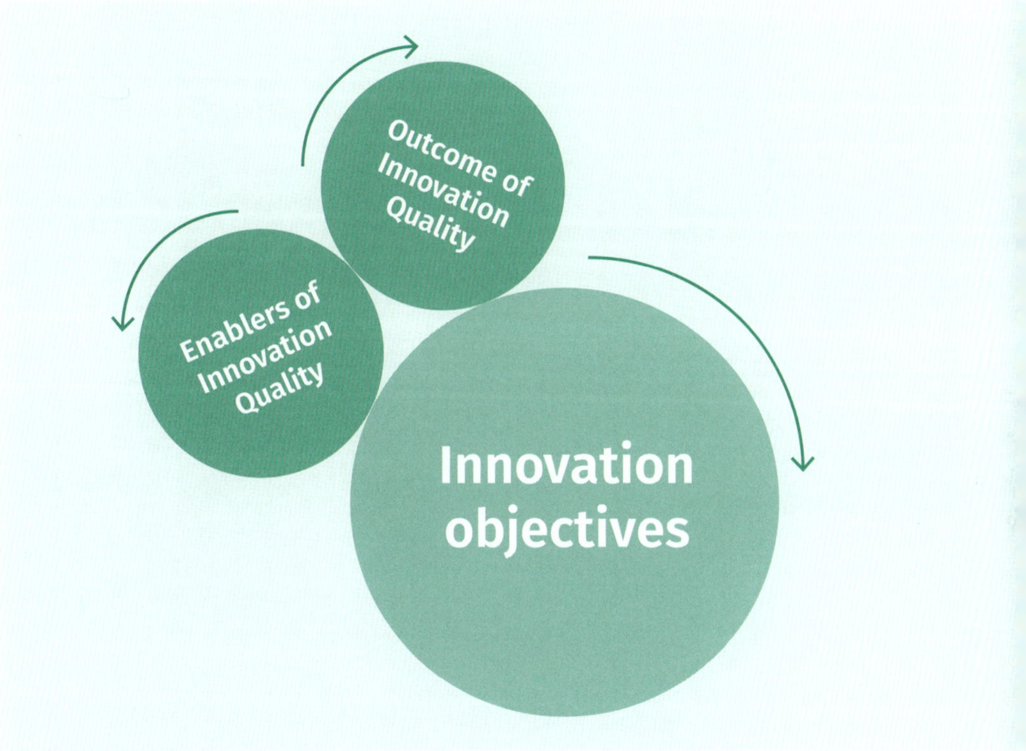

Figure 26: The laying down of innovation objectives as a prerequisite of InnovationQuality

What this means for a model for InnovationQuality is that compared to the enablers of InnovationQuality, innovation objectives operate on a different level, namely a preceding causal level. The first step has to be to formulate an innovation objective and – so that this does not just purely remain a declaration of intention but can actually become a reality – this also has to be established within an InnovationQuality project. Only then can the steps that relate to the "enabler" criteria be implemented: set up special project teams tasked with translating an idea into an innovation; set up internal and external partnerships for these projects; set up processes; provide these projects with the right means, etc.

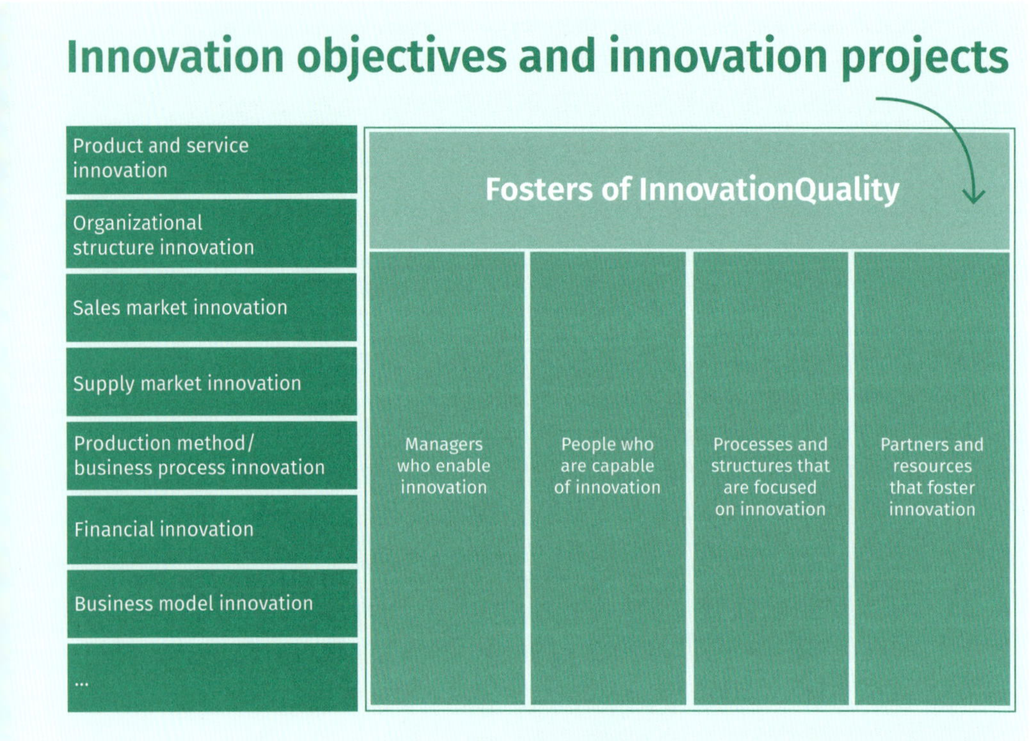

Figure 27: The preceding causal positioning of innovation objectives and innovation projects in the InnovationQuality model

2.3 The outcome of InnovationQuality

The question regarding the outcome of an innovation is for whom an innovation is ultimately of benefit. For example, an "improvement innovation" could result in a production process generating fewer rejects by introducing optimized technology; naturally, an indirect result of this could be (note: does not have to be) a reduction in end user prices and less burden on the environment due to the improved use of resources. But the actually intended and direct improvement resulting from such an incremental innovation primarily lies in the fact that a firm produces more cost-effectively since it uses resources more effectively.

Until now, thought given to the innovation value chain indicated that the very purpose of an innovation is to create value for the recipient through something new or different, and that something positive feeds back to the originator of the innovation as a result. Value should be created – or at the very least be preserved – by an innovation, for the recipient and the originator. However, the microcosm consisting of the originator and the recipient is not closed because, of course, whether it is desired or not, innovations also have an effect on stakeholders on the outside. The New also creates new risks, i.e., he who creates the New must also reckon with possibly irreversible consequences with sometimes catastrophic effects resulting from the New. The interpretation we have of value creation at this point has major overlaps with concepts such as "sustainability" or "public value." And one should be reminded at this point of lofty and fundamentally appropriate thoughts about us all being a part of this single world and that our individual actions never take place in isolation, but are forever interwoven with worldly affairs.

> *Living systems are organized in such a way that they form multi-leveled structures, each level consisting of subsystems which are wholes in regard to their parts, and parts with respect to the larger wholes. Thus molecules combine to form organelles, which in turn combine to form cells. The cells form tissues and organs, which themselves form larger systems, like the digestive system or the nervous system. These, finally, combine to form the living woman or man; and the ‚stratified order' does not end there. People form families, tribes, societies, nations. All these entities – from molecules to human beings, and on to social systems – can be regarded as wholes in the sense of being integrated structures, and also as parts of larger wholes at higher levels of complexity. [...]*
>
> *Arthur Koestler has coined the word 'holons' for these subsystems which are both wholes and parts, and he has emphasized that each holon has two opposite tendencies: an integrative tendency to function as part of the larger whole (compatibility), and a self-assertive tendency to preserve its individual autonomy (perpetuity). Biological or social systems require individuality of 'holons' to maintain the stratified order, but holons must also submit*

to the demands of the whole order to make the system viable. These two tendencies are opposite but complementary. In a healthy system – an individual, a society, or an ecosystem – there is a balance between integration and self-assertion. This balance is not static but consists of a dynamic interplay between the two complementary tendencies, which makes the whole system flexible and open to change. (Capra 1992: 38 ff.)

The word "quality" has its origins in Latin ("qualis" = of what nature/kind) and in its general interpretation, it describes the "nature," "goodness" or even "value" of an object. From the standpoint of the user, the value of an innovation indicates whether an innovation is of value for a subject in the fulfillment of a purpose. This terminological assessment of the word already explains the fundamental direction of the interpretation. However, it is not yet possible to come to a conclusion as to which form this assessment of quality – or value – takes. In the EFQM model these forms would be: people (working for the company), customers, society and the organization itself. These can provide a starting point for determining InnovationQuality, but their definition needs to be built upon from a systems theory perspective. Given this, there are five dimensions of value creation or five forms, which show the extent of the value, and thus the benefit, and thus also the quality of an innovation:

1. Value creation as the fulfillment of a holistic, system-related purpose: InnovationQuality is interpreted in value terms for a narrowly defined group of people overall (e.g., a company, an organization). InnovationQuality in terms of the outcome is reflected in whether an innovation is useful in allowing this narrowly defined community overall to achieve a desired objective. This first dimension makes immediate sense: After all, an innovation is not implemented by an organization "for the fun of it" – the expense of innovation quite simply has to pay for itself somehow. So for example, InnovationQuality from the standpoint of the innovating organization is reflected in whether and to what extent monetary or strategic objectives are achieved.

2. Value creation as the fulfillment of a granular, system-related purpose: InnovationQuality is interpreted in value terms for the members or components of a narrowly defined community (e.g., managers, workers, departments, staff, functions, etc.). InnovationQuality in terms of the outcome is reflected in whether an innovation is useful in allowing the members or components of a narrowly defined community to achieve a desired objective. A distinction

should always be made between this second dimension and the first dimension, in particular because there are sometimes spectacular differences between the assessment criteria of quality: An innovation within an organization, which is of value on the level of the overall system (e.g., through a short-term rise in productivity) can, on the people/company worker level, result in dissatisfaction or personal overload in the long term. An example of this is the introduction of the assembly line, which of course resulted in higher outputs but it reduced worker involvement to simple hand movements, thus negatively influencing work motivation.

3. Value creation as the fulfillment of a purpose of a directly environmental nature (the internal environment of the system): InnovationQuality is interpreted in value terms for a narrowly defined group of people with which an organization has a mutually existential connection ("structural coupling" – e.g., clients, shareholders). InnovationQuality in terms of the outcome is reflected in whether an innovation is useful in allowing this narrowly defined group of people to achieve its desired goals through its existential connection to the organization. This dimension also makes immediate sense, since, in the true sense, the customers of an organization should be the target of an innovation: Naturally, with an innovation relating to an organization or supply markets, the initial focus is on efficiency improvements and cost-cutting, but such innovations must always be achieved with the underlying thought that, as a result, something like a product or service will (at the very least) not become worse for the customer, so the value created for the customers is the same, or becomes even better.

4. Value creation as the fulfillment of a purpose of an indirectly environmental nature (the external environment of the system): InnovationQuality is interpreted in value terms for a group of people that has not become involved in a mutually existential connection with the organization, although it is actually affected by the actions of that organization (e.g., stakeholders, but also other enterprises and organizations). Thus assessment of the value and, with this, the quality of an innovation can and must not be made solely by a small group of customers. An innovation can of course also have implications that extend far beyond the narrow circle of the company and customers. The value and, with this, the quality of an innovation are thus also reflected in whether and to what extent this innovation affects others, even organizations outside the sector of industry. The quality of an innovation is thus also a reflection of whether

it influences the existence and survival of other organizations – for example, whether and to what extent the innovations of firms like Google and Apple have an influence on all kinds of companies and branches of industry.

5. Value creation as the fulfillment of a purpose relating to the world (the world in which the system exists and operates): InnovationQuality is interpreted in value terms for a wide- or widest-ranging group of people. The value of something is reflected in whether it is useful to a "society" as a whole, or humanity – meaning today's but also tomorrow's generations – in achieving a desired objective. On this level, we are actually dealing with something that is highly complicated and complex. Ultimately, this question leads to another question regarding value creation on this level: Has an innovation actually moved humanity forward, i.e., has life on this planet improved as a result of an innovation or is it just different? What is the use of the very latest technology or a system, if producing it or disposing of it destroys our planet in the long term? What is the use of digitization and the "Internet of Things" if this threatens countless numbers of jobs in the long term – including in science – and we have no economic or societal model with which to deal with the millions of unemployed workers?

In summary: An innovation is an idea that becomes a value-creating reality. "Becoming a reality" means that something was not merely thought about or developed into a prototype, but was actually introduced to a community. To "create" means that, in one way or another, something is new to a community and delivers value. The ultimate measure by which the value and with this the quality of an innovation can be judged, is thus the actual value creation, the actual value relating to benefit. Value creation, i.e., the evaluation of the impact of value, takes place on five levels. A distinction must therefore be made with InnovationQuality, in terms of the impact:

1. Whether and to what extent value creation results for a system as a whole (an enterprise, an organization)

2. Whether and to what extent value creation results for the members and components of a system (people working for a company, units within an organization)

3. Whether and to what extent value creation results for the direct environment of a system (customers, shareholders)

4. Whether and to what extent value creation results for the indirect environment (stakeholders, other organizations)

5. Whether and to what extent value creation results for the world as a whole in the long term.

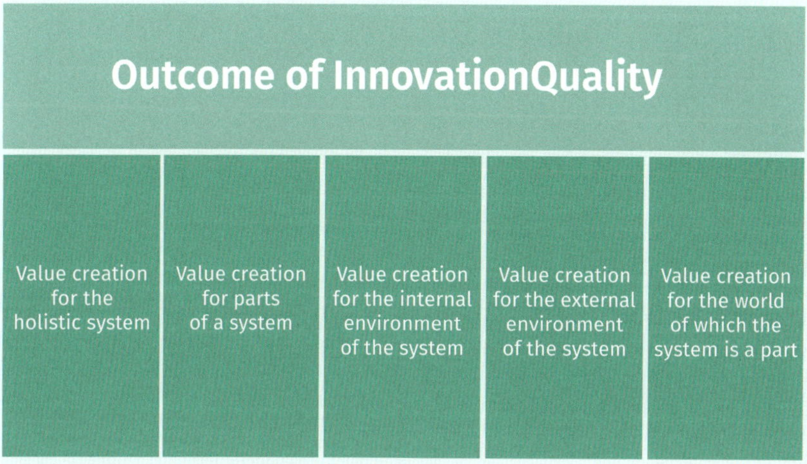

Figure 28: The outcome of InnovationQuality

Conclusion: an integrated model of InnovationQuality 3

The following provides a summary of our model. The arrows show a quasi-causal connection: Innovation objectives and the innovation projects derived from these objectives are the cause and principle for influencing InnovationQuality. The enablers are "means" in a causal context for influencing InnovationQuality; outcomes are the impact of influencing InnovationQuality.

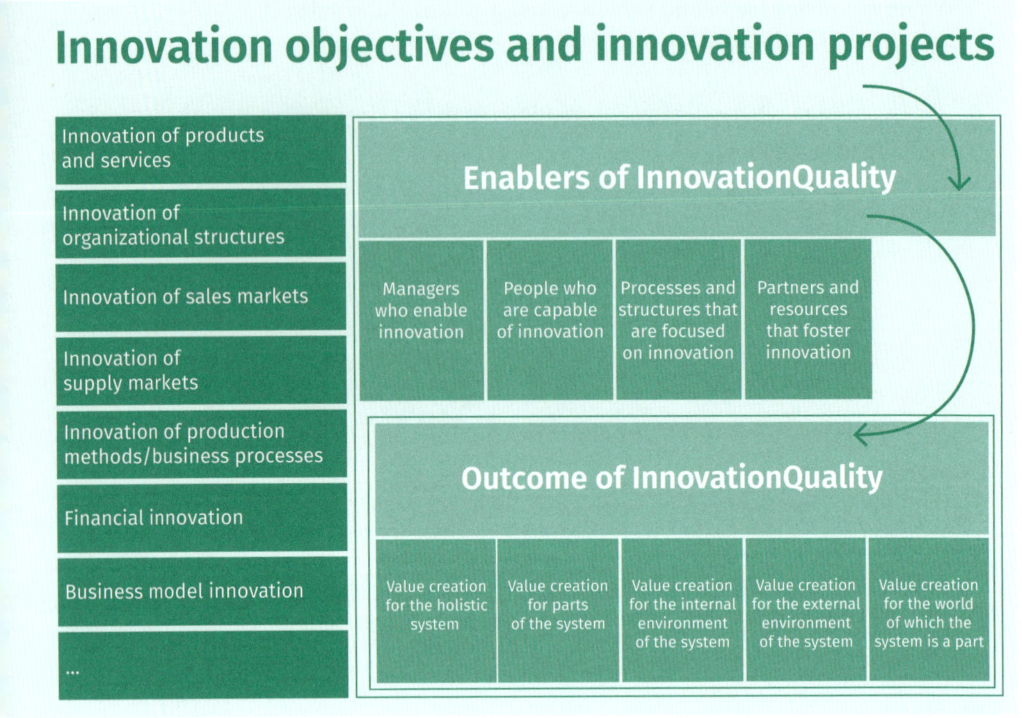

Figure 29: A model for InnovationQuality

Obviously, a model should depict something as concisely as possible. The phenomenon of "innovation" – as should be totally apparent by now – is a complex topic. A model for the quality of innovation should, like any other kind of model, attempt to simplify a complex topic. In science, there are differing opinions in this respect as to whether creating a model brings one closer to something or is actually a sin with the opposite affect.

We are aware of this conflict and admit – with complete frankness and humility – that the aim of our model would not be to somehow or other tame the diametrically opposed twin phenomena of "innovation" and "quality," united in a word called "InnovationQuality." Innovation loses nothing of its wildness, its individualism or its rebelliousness in our model. It would not only be presumptuous but also delusional if we were to believe that our model would be providing a fixed framework around the "partial phenomenon" of "innovation" just by introducing the phenomenon of "quality" and that this framework would make it possible to look at the phenomenon from a central perspective. Our model does nothing to tame the phenomenon of "innovation," it does nothing to bring anything under better control. But we do hope our term InnovationQuality and the model for InnovationQuality provoke discussion, not only about our theories but also about the concept of "InnovationQuality."

Step 1: Introduction to the two terms of “Quality” and “Innovation” by closely examining the phenomena and their distinctive characteristics

Step 2: Combination of the two terms, or the two phenomena of “innovation” and “quality”

Step 3: Specific definition of the interpretation of “ InnovationQuality”, also as a quantitative variable

Step 4: Outline of a model for InnovationQuality

Step 5:

Illustration of an approach for managing InnovationQuality

In the final chapter we introduce a concept in five steps. These make it possible to manage InnovationQuality. We use the terms "innovation" and "InnovationQuality" to refer to a variety of phenomena. What all of these phenomena have in common is that they represent something new in one way or another. In doing so, it does not matter if the New is something incremental or radically new. It also does not matter if the New is new for its originator or the recipient (novelty through diffusion or novelty through adoption). It also does not matter if the New has an evolutionary or a disruptive impact. And finally, it does not matter if the New is a new engineering technique, a new technology, a new type of organization, a new way of doing business or something else which is new.

However, there is a consequence of this broad interpretation of the terms "innovation" and "InnovationQuality": Conventional concepts of "innovation management" look primarily or even exclusively at innovation of the "new product" variety. For a broad interpretation of innovation and, with this, InnovationQuality, one thus needs a different, essentially heuristic concept for managing innovations and in particular for managing InnovationQuality. However, this concept does not merely serve as a management concept. Instead, such a heuristic approach provides an opportunity to examine specific and contextual factors dictating the success or failure of an innovation. In other words, the following heuristic offers a paradigmatic entry point, not only for implementation in enterprises but also for the scientific research of innovation and its quality.

The strategic triangle of business development 1

There are of course a not inconsiderable number of models on the management of innovation. One of the weaknesses of these models is that they are often based on a narrow interpretation of innovation – particularly product innovation. The aim of the following is thus to introduce the heuristic of the "strategic triangle of business development." The "strategic triangle of business development" is basically open and a suitable starting point for all kinds of development processes – including, importantly, areas beyond the realms of business.

The philosopher Odo Marquard said that "the future needs the past" (Siemens AG 1994). Schumpeter also emphasized the fact that every specific business development process is based on previous developments, that even the most enthusiastic entrepreneur has to use the existing facts relating to a company and derive his decisions from these, and that the future can only create something from which a foundation has already been laid for today (Schumpeter 1934). Business development is therefore a process over time, caught in a conflict between the demands and the possibilities of the internal and external environment of the company; in other words, the company's past, the company's present and the company's future. Accordingly, business development is only possible if future plans for the company are based on the consideration of origins, i.e., the past and the present. Only this way can a process be put in place for a company to learn, providing a basis for entrepreneurial growth (Bleicher, 2004).

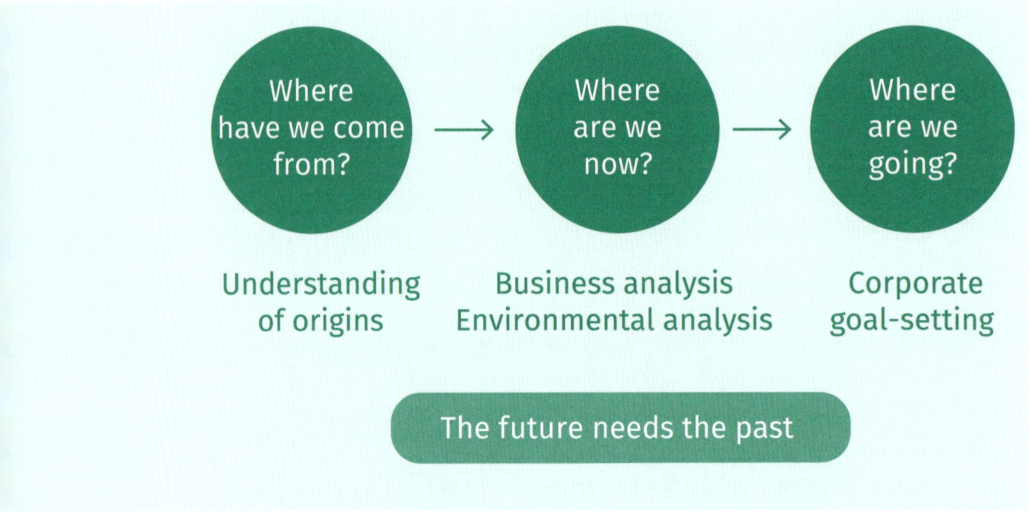

Figure 30: Business development in a temporal and causal context

Another thing this illustration shows is that corporate development is a dynamic, progressive process at all times. This dynamic process lays a foundation for four aspects of business development (Faix, Buchwald, Wetzler 1994; Rasner, Füser, Faix 1999; Faix, Rasner, Schuch 1996):

1. The current situation, i.e., the situation faced by the company at the moment

2. The framework conditions under which the company operates

3. The target situation, i.e., the objectives of the company for the future

4. The path from the given, current situation faced by the company to the target situation of the company, i.e., to achieve the corporate objectives. This path is the strategy of the company.

The following illustration was introduced by Faix et al. as the "strategic triangle of the transformation process" and can, from today's point of view, be called the "strategic triangle of business development."

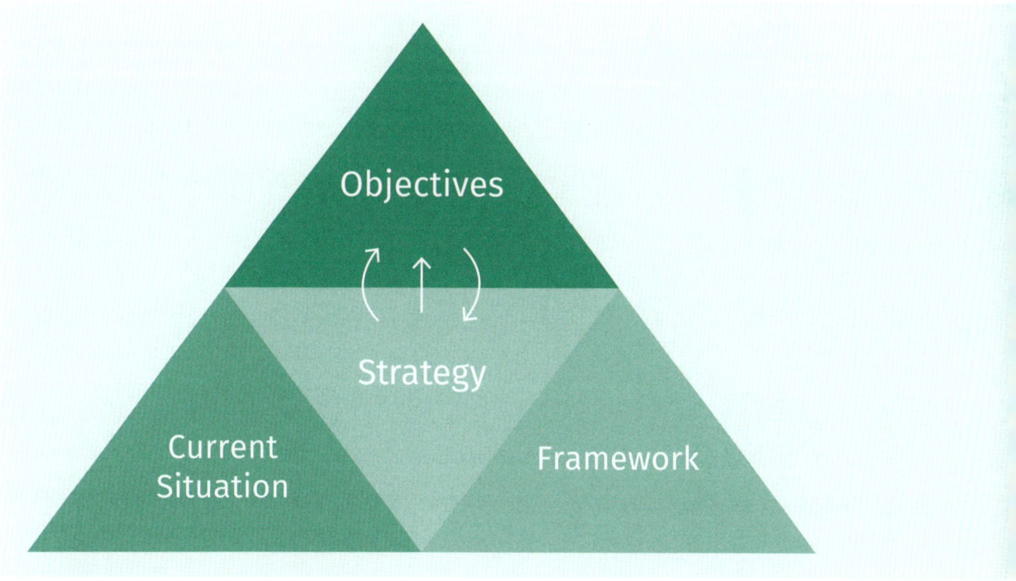

Figure 31: The strategic triangle of business and project development (Faix et al. 2008)

The business development process overall spans the following eight steps, which continually rewind.

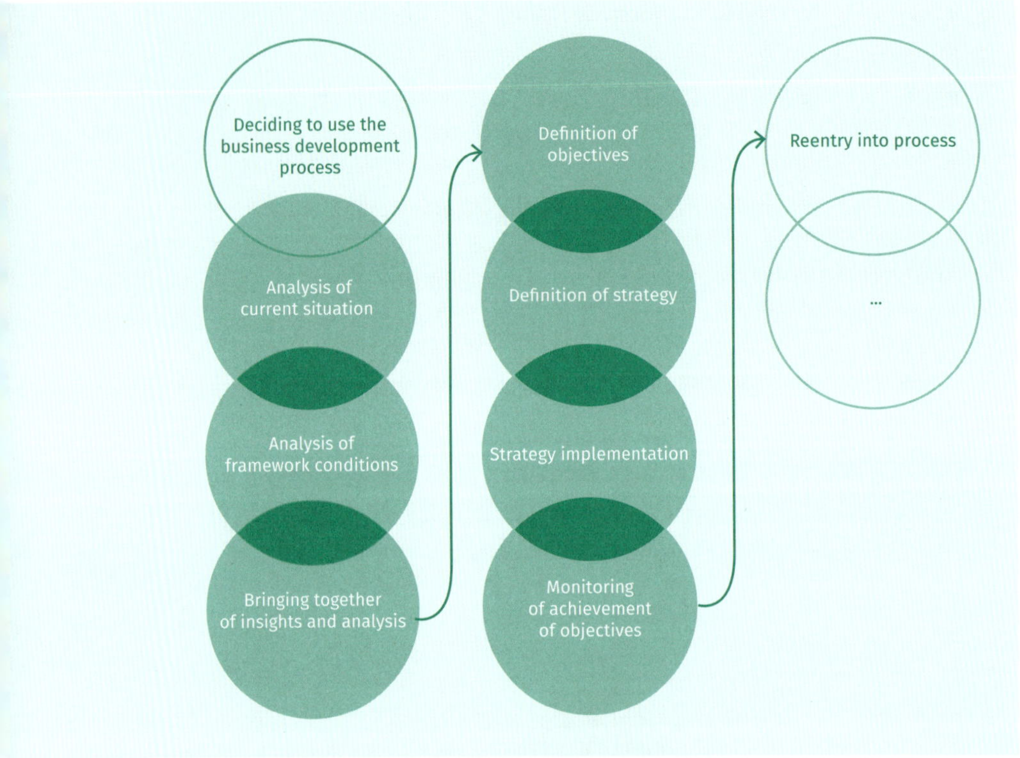

Figure 32: The development process of a company or a project

Phase 1	Decision to use the development process – a company must make a conscious decision to develop as a business
Phase 2a	Analysis of the current situation affecting the company/project
Phase 2b	Analysis of framework conditions affecting the company/project
Phase 3	Based on results of the analysis, definition and evaluation of opportunities and threats affecting the company/project
Phase 4	Definition of company objectives or project objectives. Once the company or project objectives have been formulated, objectives can be broken down into objectives within individual areas of the company hierarchy or sub-objectives for the project.
Phase 5	Definition of the strategy, i.e., the plan for achieving the company or project objectives. This also involves first developing a general strategy and then breaking this down into sub-strategies for individual areas of the company hierarchy or sub-objectives for the project.
Phase 6	Implementation of the strategy at all levels of the company hierarchy or implementation of all sub-objectives for the project
Phase 7	Monitoring of achievement of objectives, i.e., has what was defined as the objective been achieved exactly, and, if not, what are the differences, shortfalls, or both
Phase 8	Entering back into the process – given the need to achieve business growth, it is essential to keep developing

Table 4: The development process of a company or project

2 The strategic triangle of innovation and InnovationQuality management

The aim of the following is to specifically adapt the heuristics of the strategic triangle and the steps that are derived from the triangle to managing innovation and InnovationQuality. The resulting process also spans eight steps:

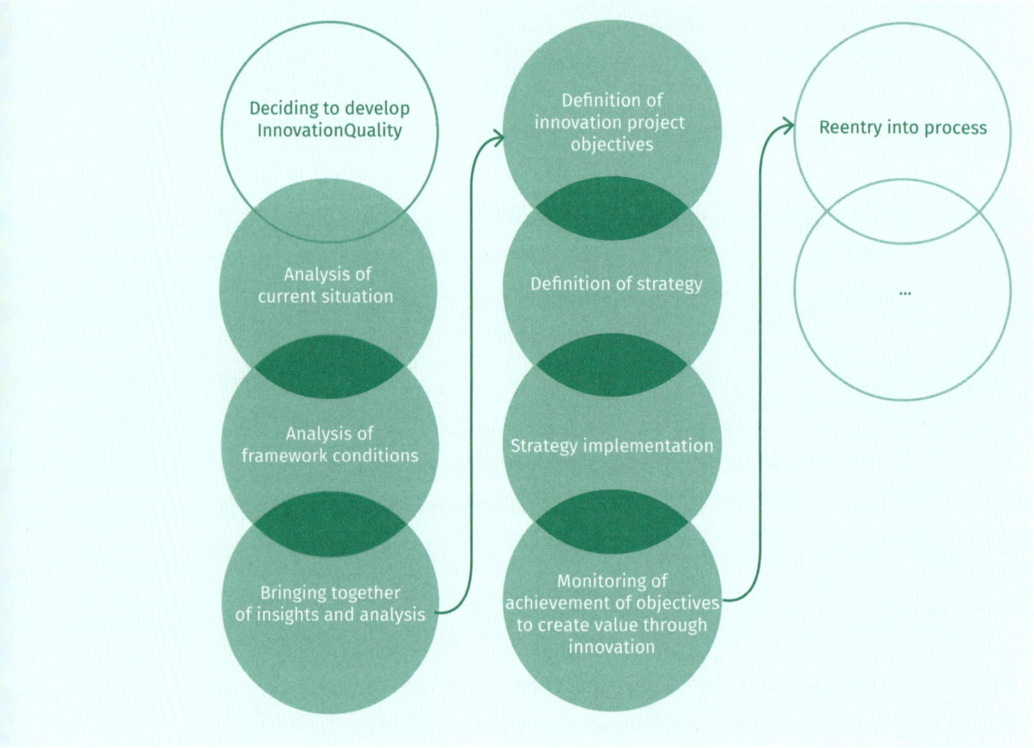

Figure 33: The development process of InnovationQuality

Phase 1	Decision to develop InnovationQuality
Phase 2a	Analysis of the current situation with respect to current and possible value creation
Phase 2b	Analysis of framework conditions with respect to current and possible value creation
Phase 3	Definition and evaluation of opportunities and threats
Phase 4	Definition of specific innovation project objectives
Phase 5	Definition of strategy
Phase 6	Implementation of strategy
Phase 7	Monitoring of achievement of objectives with respect to value creation through innovation
Phase 8	Entering back into the process

Table 5: The development process of InnovationQuality

2.1 Phase 1: Decision to develop InnovationQuality

One of the first and most important steps is deciding in favor of developing InnovationQuality in the first place; that is, defining innovation objectives in general, and, in particular, defining InnovationQuality objectives. In other words: achieving tangibly expected value creation by realizing innovation objectives. To ensure these objectives do not just remain general statements of intention, as soon as the decision has been made, corresponding innovation projects should be initiated.

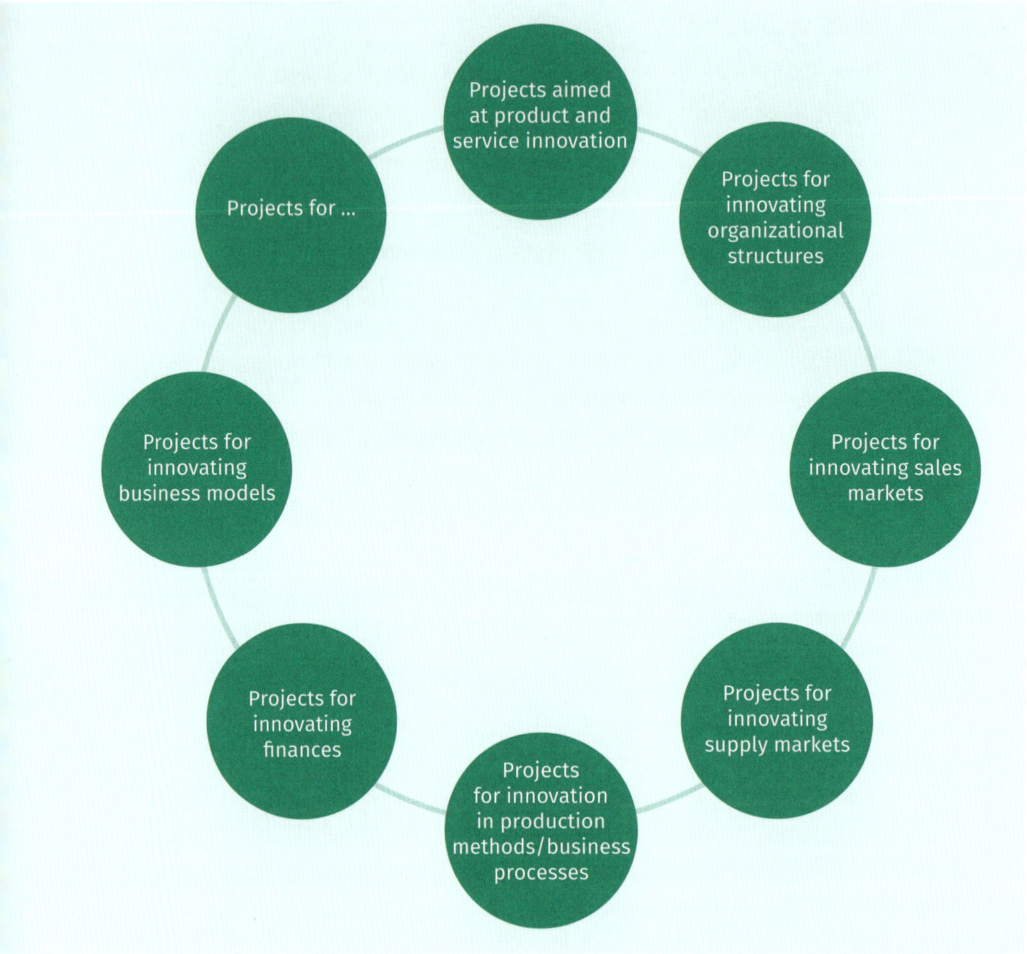

Figure 34: The initiation of innovation projects as the first key step in realizing innovation objectives and developing InnovationQuality

The aforementioned "innovation objectives" and "InnovationQuality objectives" should not be confused with the specific "innovation project objectives" that arise in the next step. The reasons for formulating and declaring these objectives is less about being able to start innovating immediately. According to scientific theory, each of these objectives relates to a problem area to a certain degree, which is made tangible and serves a purpose at a later stage, once concrete questions have been posed and objectives are set. Innovation objectives thus serve less of a pragmatic and much more of a paradigmatic purpose, since it is only once it has been decided that something should be researched at all (along with what must be done and why) that the following steps can be given a basic structure and actually become possible and necessary. Innovation objectives and InnovationQuality objectives provide that certain first impetus as one's awareness is directed toward something specific. And it is precisely here that it is decisive whether one has a broad or narrow interpretation of "innovation" and, with this, one is aware of the possible sources of value creation. If one only sees innovation within the context of "new products," one will only think in this direction in the following steps, i.e., conduct analysis relating to new products, lay down objectives relating to new products, and formulate and implement strategies relating to new products. If, however, one interprets innovation more broadly, one broadens one's awareness to the many possible ways one can create value with the New – and one will also perceive these possibilities within the company and respond to them.

2.2 Phase 2: Analysis of the current situation and the framework conditions with respect to current and possible value creation

It is not possible to define the specific results that should be achieved with an innovation project (which can include the InnovationQuality to be developed as a result of the project) and, with this, the corresponding strategy (possible ways to achieve innovation project objectives) unless the current situation (the status of the company or the project) and the framework conditions (changes in the company's environment) are understood, and that all these things have been summarized into assumptions. This process of analysis and synthesis should be extremely thorough as all subsequent decisions depend on it. So it can be extremely useful to leave existing behaviors and opinions behind and consider the existing company and its competitive field from different angles.[16] A more or less exact and meaningful picture of the present situation faced by the company can be arrived at by analyzing factors within the company and comparing this to similar evaluations carried out on the company in the past.

An analysis of the current situation primarily covers the following aspects: To determine the change in value creation achieved with an innovation at a

16 If somebody new looks at the picture, for example, if someone from outside the company (e.g., an external consultant) or a new employee is asked to carry out the task, the result is almost inevitably a new and more objective view of the company and its competitive field.

later point, the outcome of current value creation activities should be analyzed. This analysis should be carried out with respect to the innovation project and the company itself. The first key question to be posed when analyzing the current situation is therefore: What value is currently created directly in the area in which the innovation is to be realized? And what is the share of this value creation in relation to the overall company? For a company, value creation through innovation means organic growth – growth through one's own energy and "one's own physical means." To define a project later on, which is both ambitious and realistic, the current situation regarding the "enablers" of innovation should be captured. The second key question when analyzing the current situation is therefore: What is the current situation regarding

> the innovative capability of employees
> the innovation enablement of managers
> the innovation orientation of processes and structures
> the innovation support of partners and resources?

Tried and tested methods and tools that are frequently used for the analysis of this current situation are:

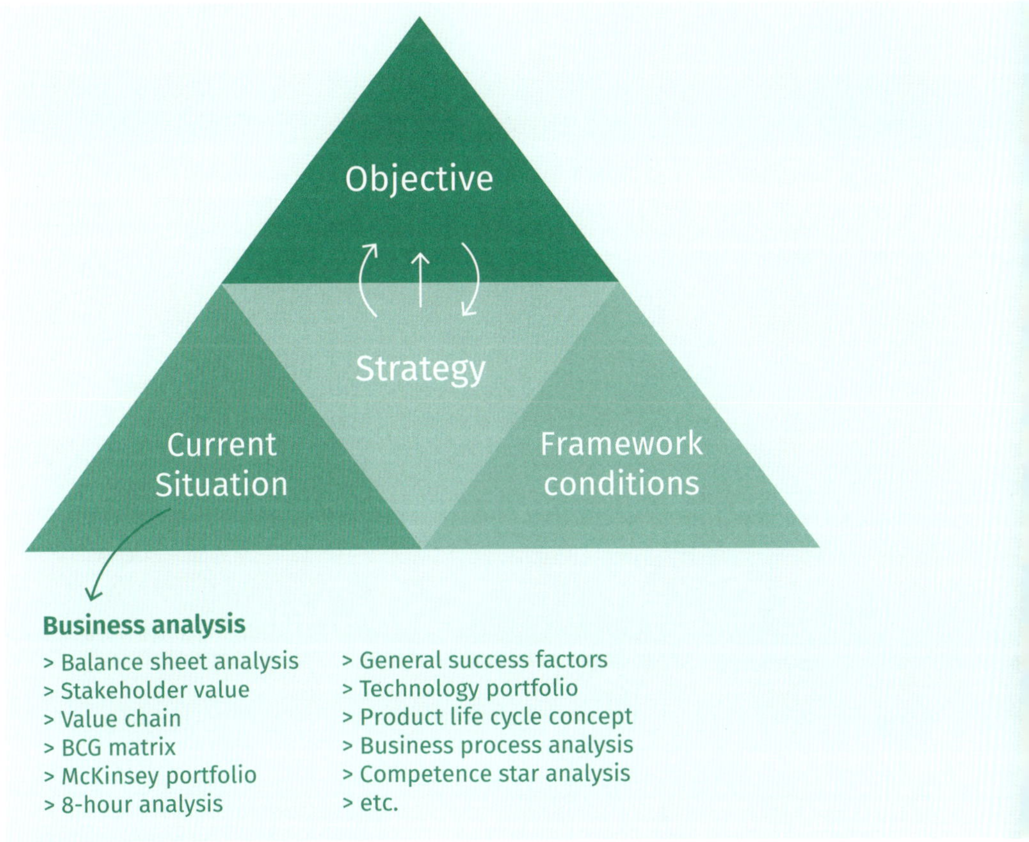

Figure 35: Analysis of the current situation

On the one hand, an analysis of the framework conditions should include comprehensive monitoring of the current endogenic situation. This includes factors such as an analysis of competitive forces and a market analysis, as well as an assessment of the political, macro-economic, social, technological, legal and environmental conditions. On the other hand, analyzing the framework conditions involves thinking in detail about the future, i.e. primarily: Which megatrends are currently happening and which future customer requirements can be expected to develop as a result of this? Tried and tested methods and tools that are frequently used for the analysis of the framework conditions are:

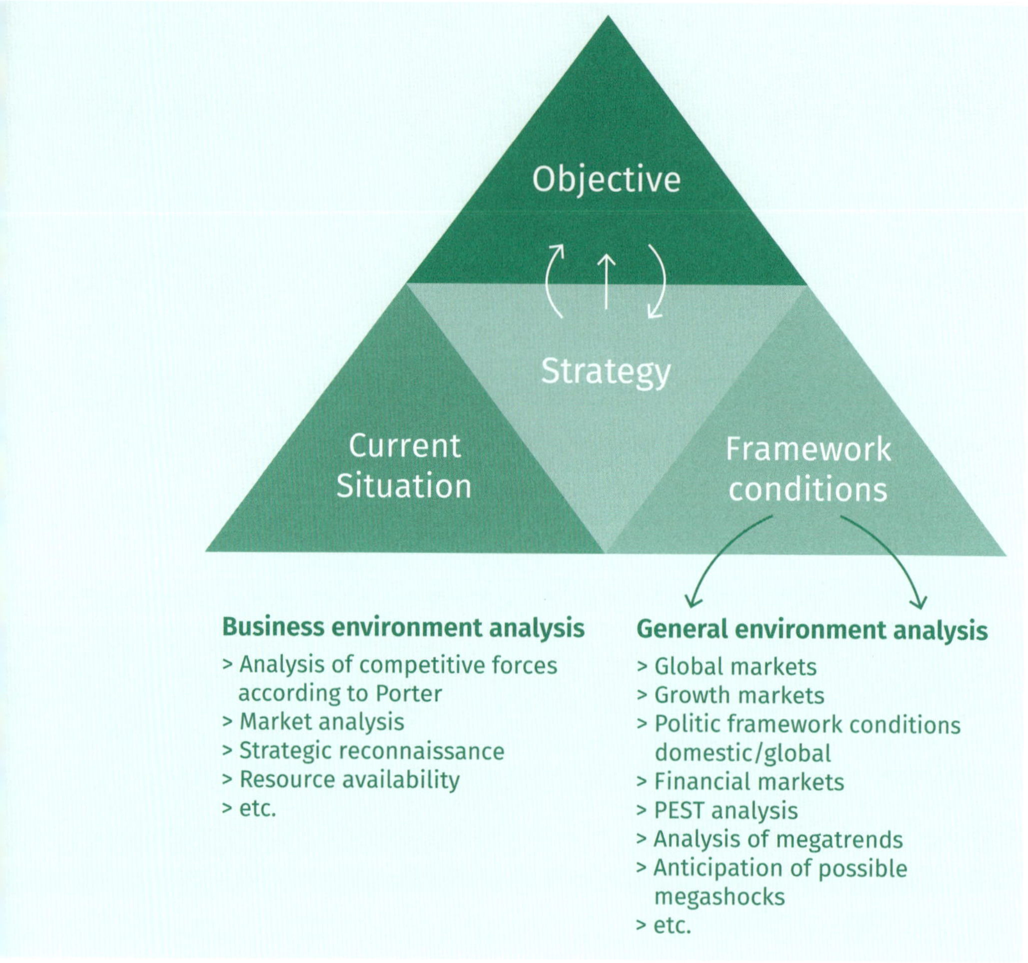

Figure 36: Analysis of framework conditions

2.3 Phase 3: Definition and evaluation of opportunities and threats

Merging the analysis of the current situation with the analysis of the framework conditions makes it possible to derive so-called strategic concepts, i.e., a synthesis of both analyses shows the potential for and threats to the current and future value creation of a company. Strategic concepts are more of a strong predictive nature than merely analytical tools. This systematic assessment of the future makes it possible to come to an initial conclusion regarding the performance indicator "PIn" – the probability of occurrence, or the likelihood of value creation as a result of innovation.

Tried and tested methods and tools that are frequently used for the definition and evaluation of strategic concepts are:

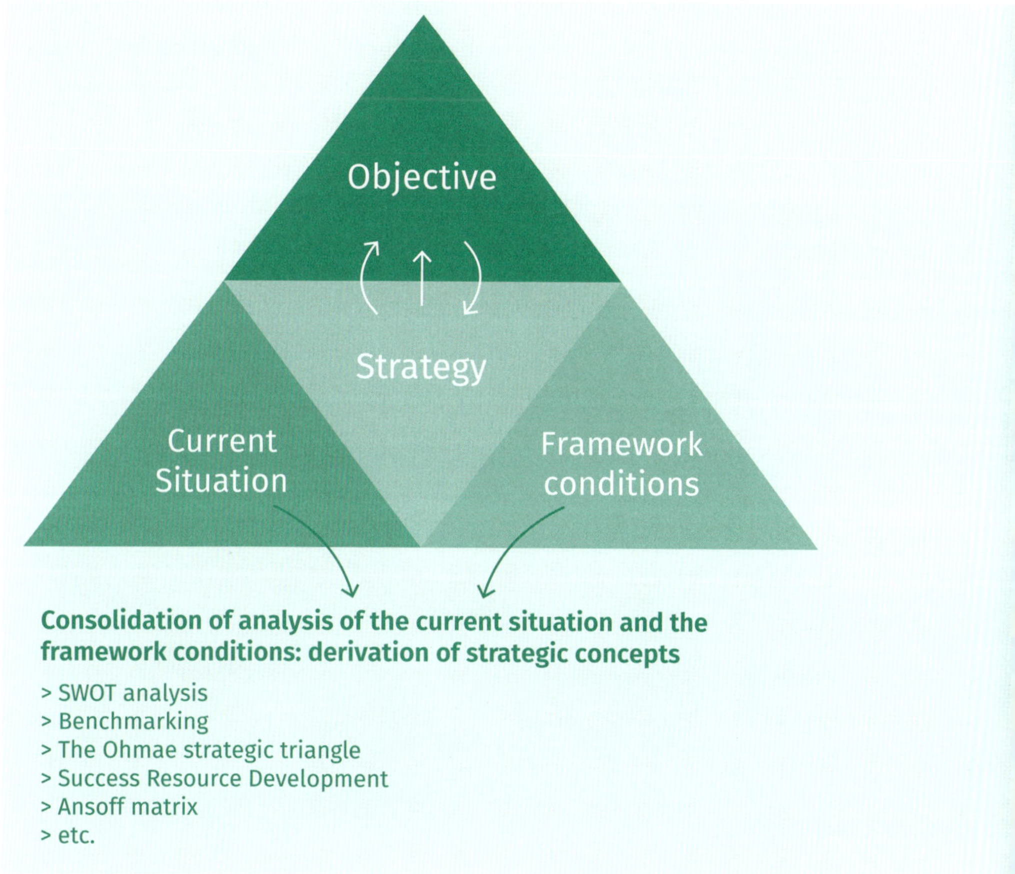

Figure 37: Derivation of strategic concepts

2.4 Phase 4: Definition of specific innovation project outcomes (InnovationQuality project objectives)

The next step is to combine the results of the analysis of the current situation and the framework conditions, including a synthesis of this in the form of strategic concepts, to form a holistic process. This simultaneously makes it possible to derive "innovative" and realistic outcomes, i.e., innovation project objectives and, in particular, objectives for InnovationQuality.

An innovation project objective spans the following dimensions:

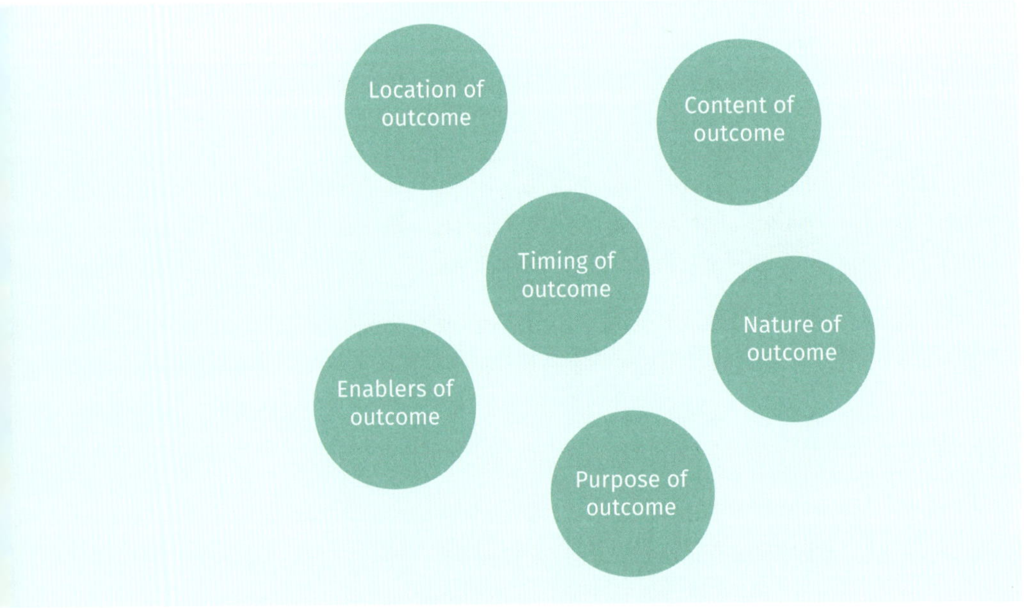

Figure 38: Dimensions of InnovationQuality project objectives

The content of the outcome reflects the specific nature of the New, actually emanating from the world of the originator into the world of the recipient (e.g., customers). To repeat, according to Schumpeter, something new results when existing things are combined in a new or a not yet implemented way – or as Schumpeter actually expresses it: An innovation is the outcome of the combination of factors and "the carrying out of new combinations." (Schumpeter 1934: 65-66)

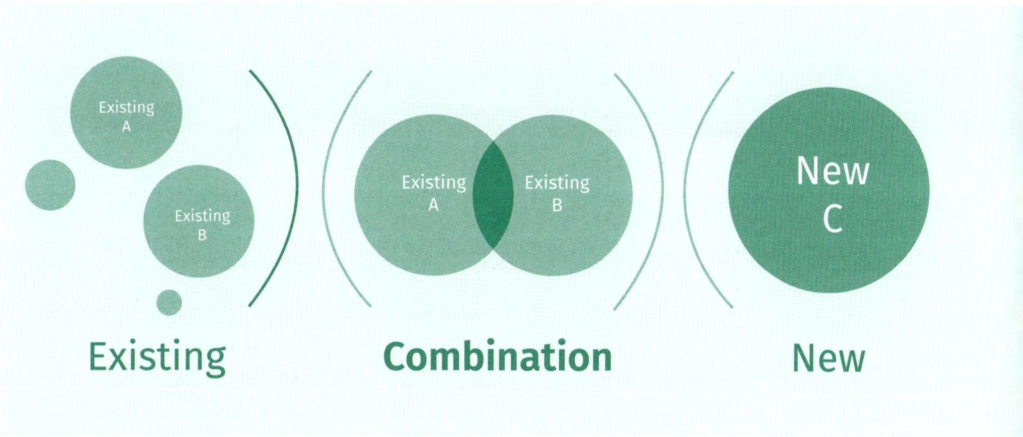

Figure 39: The New as a combination of the existing

Thus, to define the innovative content of an objective, one needs to do nothing more than combine the existing. What sounds so simple – "combination of the existing" – is actually a process than can be critical and difficult to influence. The process can be critical because it is only when this step happens that an innovation objective can be arrived at, which is sufficiently detailed for the situation facing the company. The process can be difficult to influence because one is not dealing with technocratic procedures at this stage but instead with actions that are inventive, artistic and creative. Despite this, establishing innovation objectives can of course be systematically conducted and enabled.

As already explained in the introduction, the principle underlying "the New as a combination of the existing" is not only witnessed in the world of business but also in the language we speak, namely in the form of metaphors. A metaphor is an expression whereby a term or group of terms are combined with other terms or groups of terms. According to (structural) semantics, a term involves a group of distinctive features. A metaphor, or a combination of terms, is the result of the features of Term A being transferred to Term B. To create or understand a metaphor, one needs knowledge of certain things:

> Knowledge regarding the distinctive characteristics of Term A
> Knowledge regarding the distinctive characteristics of Term B.

Only with such knowledge can characteristics be transferred and, with this, a connection be made between terms.

For the innovation process examined here, the previously outlined analyses provide a basis for this combination.

> By analyzing the current situation facing the business, detailed knowledge is required regarding what the company is and what things are like for the company
> By analyzing the framework conditions, detailed knowledge is required regarding the nature of the business environment.

Together, these analyses provide distinctive characteristics that reflect the current "essence" of the company or project and its environment. In-depth, comprehensive knowledge of these distinctive characteristics provides a basis of combination and, with this, the innovation objections.

In turn, by deriving strategic concepts, these two analyses can be aggregated into a construct of a higher order. This construct can also be combined with other "things," i.e., with insights from other areas of science. A prime example of this approach is bionics, where knowledge of biology and zoology is combined with (technological) science.

According to the theory of metaphors, the less that combined objects appear to have in common (at first glance), the more original the metaphor. In a similar way, it can be assumed that the less the existing "raw material of knowledge" has in common, the "more radical" the innovation objective is. As a result, establishing innovation objectives is, on the one hand, primarily about systematically thinking laterally and differently, and seeing beyond the horizon. On the other hand, establishing innovation objectives is about having the courage to support supposed "fooling around" and actively encouraging it. This is because lateral thinkers are often "lateral drivers" with the fortune to have met the right people at the right time – otherwise, they could now just be "crazy idiots."

The nature of the outcome reflects what would be or would actually occur if there were feedback from the recipient for whom the innovation was intended. Unlike the previous point – the "content of the outcome" – there is a difference with the "nature of the outcome" in terms of both timing and the content. The issue is not what an innovation is, but the impact that results from it – or more specifically: the value creation. The "nature of the outcome" thus describes InnovationQuality as a target in a narrower sense, but of course this target is to a large degree incomplete if it is not specified alongside the further dimensions of the objective described in this chapter.

With the "nature of the outcome" dimension, a target value should be set for each individual project, in other words it should be defined (in terms of a minimum, optimum or ideal) how big the change in value creation should be by implementing an innovation project.

$$InQ_{Project} = \Delta cVIn_{Project}$$

Formula 16: Definition of a target value for InnovationQuality resulting from an innovation project

This means that relative and absolute value creation ($cVIn_{relative}$ and $cVIn_{absolute}$) should be defined, which should be achieved by a specific innovation project within a specific timeframe.

$$cVIn_{Project} = \frac{\text{Outcome after implementation of an innovation project}}{\text{Outcome before implementation of an innovation project}}$$

Formula 17: Value creation "cVIn" resulting from an innovation project (relative)

$$cVIn \geq 1 \text{ means value creation}$$
$$cVIn = 1 \text{ means value retention}$$
$$cVIn \leq 1 \text{ means value reduction}$$

Formula 18: Interpretation of the indicator "cVIn"

$$cVIn_{Project} = \text{Outcome after implementation of an innovation project} - \text{Outcome before implementation of an innovation project}$$

Formula 19: Value creation "cVIn" resulting from an innovation project (absolute)

To make it possible to pull together the outcome of individual innovations to arrive at Corporate InnovationQuality ("Corporate-InQ"), the value creation of individual innovations has to be monetized, that is: The value creation of an innovation should be seen as the ratio between turnover/profit/revenue/etc.

before diffusion of an innovation and the turnover/profit/revenue/etc. after diffusion of an innovation.

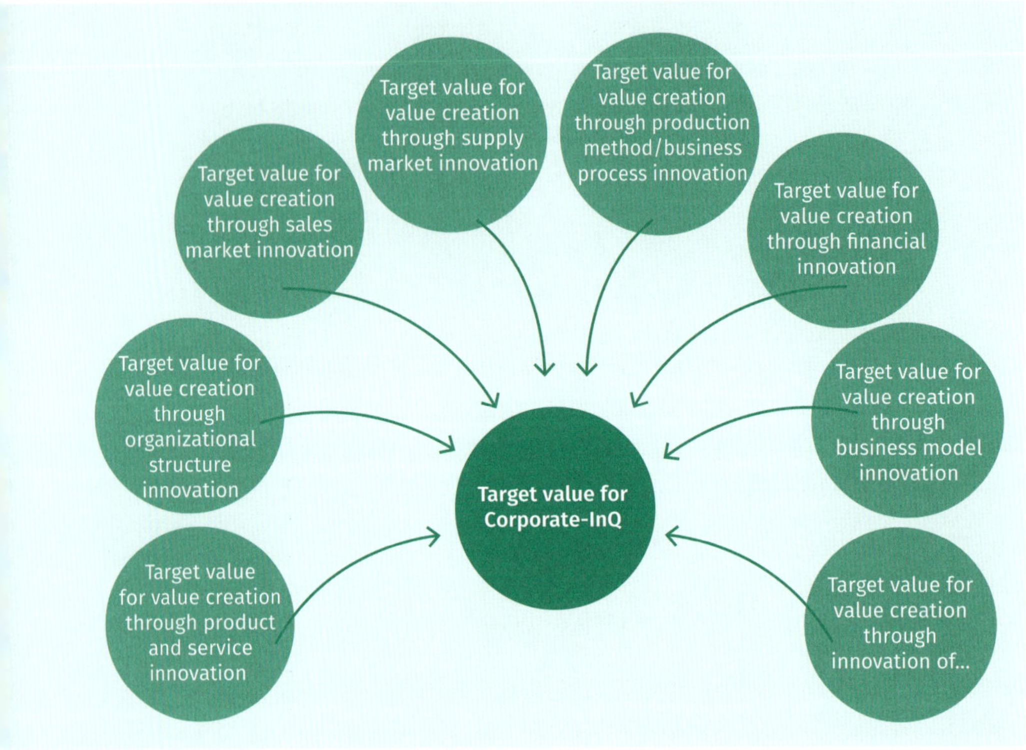

Figure 40: Target setting for value creation resulting from individual projects (InnovationQuality of individual projects) and their cumulative contribution to the overall value creation of the business (Corporate-InnovationQuality)

The purpose of the outcome reflects the company's underlying motivation for its innovation. The purpose is directly derived from the need that the company is attempting to satisfy with an innovation. In keeping with Maslow's hierarchy of needs (Maslow 1954), a pyramid of needs can also be formed for companies. This pyramid of needs works on three levels (Rasner, Füser, Faix 1999):

- Need to survive
- Need to grow and globalize
- Need to shape the future

These three levels can be seen very vividly in the New Economy. At the beginning of the Internet hype, there were numerous startups, all on the bottom level of the pyramid. These companies tried to establish their innovation or idea on the marketplace, recruit employees, acquire customers and generate turnover. Out of this broad base of companies, only a fraction worked their way up to the next level. Companies on this second level increasingly made profits and promoted the further development of the Web with innovations. Only very few companies (e.g., Google) are now at the third and highest level. Google is attempting on one hand to build upon its power and influence in numerous new ways; on the other, it is expanding its supremacy in new markets. Each company defines a different underlying motivation for innovation, depending on the level where it currently finds itself.

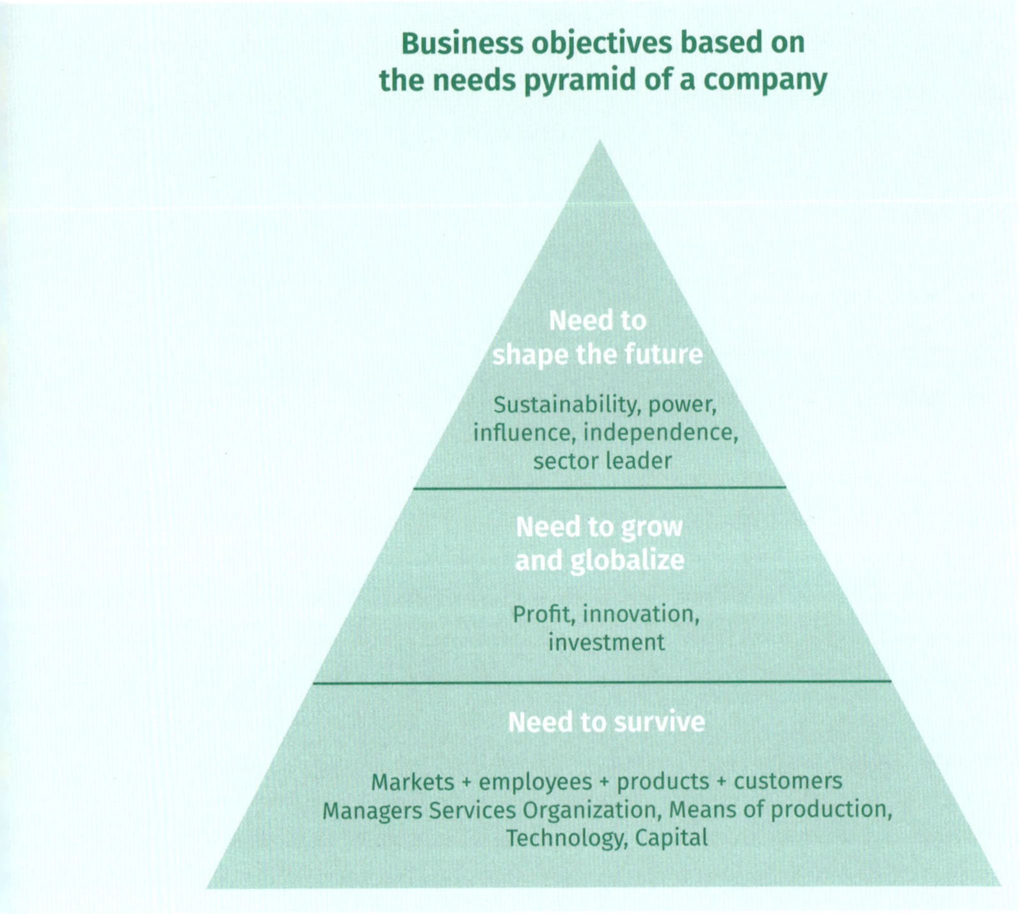

Figure 41: Motivation for "Improving InnovationQuality" expressed in terms of the actual needs of a company (Rasner, Füser, Faix 1999)

The timing of the outcome can be interpreted in a number of ways. In project management, it is commonly seen as the point at which a project is completed. Relating this back to an innovation project, this would correspond to the moment of diffusion. As we ascertained quite early on, however, actual value creation resulting from an innovation does not start until there is positive feedback from the recipient of the innovation. In short: Value creation resulting from an innovation – and thus InnovationQuality – is not denoted by whether an innovation has been shared, but instead by when the "money starts coming in." As a result, we do not want to specify the "timing" dimension as the precisely definable moment of diffusion, but rather refer to a somewhat abstract moment of feedback. This moment is abstract because a business has to decided for itself when or by when the impact of such feedback should be felt. For example, one could set a deadline, starting with the moment of diffusion such that within *x* months or years this or that value creation should be achieved. In terms of a formula for InnovationQuality, this point in time is the value t, by which time the defined value creation cVIn should result from an innovations project.

$$InQ_{t\text{-}Project} = \Delta cVIn_{t\text{-}Project}$$

Formula 20: The timing of value creation resulting from an innovation project

The location of the outcome of an innovation project describes the sphere of influence in which value creation is initiated, safeguarded or sustained as a result of an innovation. Insofar as it is possible and makes sense to arrive at a holistic definition of the outcome of InnovationQuality, the outcomes for all possible spheres of influence of an innovation should be captured. In other words, at this point, one should consider and formulate whether and to what extent:

- value creation results for a system as a hole (a business, an organization)
- value creation results for the members and constituent parts of a system
- (employees, organizational units)
- value creation results for the immediate environment of a system (clients, shareholders)
- value creation results for the indirect environment (stakeholders, other organization)
- value creation results for the world as a whole in the long term.

The enablers of the outcome dimension reflects the fact that the realization of any innovation depends on the active involvement of people. As a result, when defining the objectives of an innovation project, specific reference should be made to the "enablers" of an innovation, i.e., what involvement should there be of people who are capable of innovation, of managers who enable innovation, of processes and structures that are focused on innovation, and of partners and resources that foster innovation, in order for an innovation project to be ultimately implemented?

2.5 Phase 5 and 6: Definition and implementation of the strategy

The next step is to define a strategy and this involves finally finding a suitable way to achieve the identified innovation project objectives in light of the known conditions (current situation, framework conditions). Although defining a strategy is naturally not the same as implementing it, we would still like to pull both of these steps together at this point. This is because our aim now is to underscore once again that an innovation as Schumpeter described it is not the "defining of new things" but instead the "doing of new things." Naturally there are huge differences in terms of content between strategies and how innovation projects are implemented, depending on the type of innovation (e.g., new product, new sales market, new business model, etc.). Nevertheless, on a formal level, all innovation projects adhere to a similar logic:

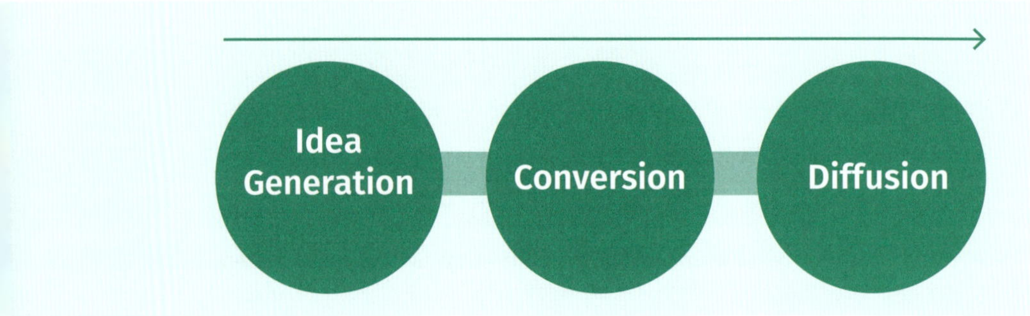

Figure 42: The steps followed by an innovation

When actually planning a project, the following milestones can be defined and set.

Milestone: X valuable ideas	in a department, in a function of a business	Idea generation
	through collaboration between departments/functions of a business	
	through collaboration between a business and external partners	
Milestone: X ideas that make it beyond "selection" and "initial funding	Selection (screening and startup funding, or, in general: provision of means, Initial funding)	Conversion
Milestone: Point in time X to first diffusion (early adoption)	Development (from initial idea to first outcome)	
Milestone: Point in time X to complete diffusion	Vermarktung oder allgemein: Verbreitung von Ergebnissen der Umsetzung	Diffusion

Figure 43: Milestone setting within the context of the definition and implementation of innovation projects; based on the idea of the "Innovation Value Chain", Morten T. Hansen and Julian Birkinshaw (2007)

2.6 Phase 7: Monitoring of achievement of objectives regarding value creation resulting from an innovation

Of course, how long an innovation project lasts or the time horizons of innovation can vary tremendously. There can be a number of months if not years between the initial starting point of an innovation and ultimate completion of an innovation project. Nevertheless, it is necessary to assess or monitor value creation after diffusion of an innovation and, with this, also the quality of the innovation. In doing so, the timing of monitoring may be different, depending on the specific situation: Value creation that can result from a product innovation tends to be perceived earlier than the value creation that can result from an organizational innovation. In principle, one could consider this step the "moment of truth" for InnovationQuality since it is now that it becomes clear whether and to what extent an innovation is actually of value.

As a starting point for assessing InnovationQuality, all formula relating to InnovationQuality can be revisited.

$$InQ_t = \Delta cVIn_t$$

Formula 21: InnovationQuality "InQ"

$$cVIn_{relative} = \frac{\text{Outcome after diffusion of an innovation}}{\text{Outcome before diffusion of an innovation}}$$

Formula 22: Value creation "cVIn" (relative)

$$cVIn \geq 1 \text{ means value creation}$$
$$cVIn = 1 \text{ means value retention}$$
$$cVIn \leq 1 \text{ means value reduction}$$

Formula 23: Interpretation of the indicator "cVIn"

$$cVIn_{absolute} = \text{Outcome after diffusion of an innovation} - \text{Outcome before diffusion of an innovation}$$

Formula 24: Value creation "cVIn" (absolute)

$$cVIn^{*}_{relative} = \frac{\textit{Turnover after diffusion of an innovation}}{\textit{Turnover before diffusion of an innovation}}$$

Formula 25: cVIn (relative) within the context of the business indicator "turnover"

$$cVIn^{*}_{absolute} = \textit{Turnover after diffusion of an innovation} - \textit{Turnover before diffusion of an innovation}$$

Formula 26: cVIn (absolute) within the context of the business indicator "turnover"

$$cVIn^{**}_{relative} = \frac{\textit{Profit after diffusion of an innovation}}{\textit{Profit before diffusion of an innovation}}$$

Formula 27: cVIn (relative) within the context of the business indicator "profit"

$$cVIn^{**}_{absolute} = \textit{Profit after diffusion of an innovation} - \textit{Profit before diffusion of an innovation}$$

Formula 28: cVIn (absolute) within the context of the business indicator "profit"

Phase 8: Entering back into the process 2.7

Companies have little choice but to consciously revitalize and reorganize their value creation in a systematic and sustainable manner. Without this, it is not possible for them to improve InnovationQuality and thus safeguard their chances of survival and long-term success. Companies have to be conscious of the "tortoise and the hare" principle: A company has to keep ahead of the first copycats of the company's products or services in terms of quality, quantity and/or time. Continuously maintaining this distance in front of the first copycat is actually not just about product or service innovation: It is also possible to copy production methods, business processes, strategies used in sales and supply markets, or more effective organizational structures. It is only possible for a company to maintain and extend its competitiveness by innovating on all levels and in all areas of the business, thus always staying one step ahead of adversaries. Accordingly, innovation is a never-ending development routine that must be continuously followed by the whole business.

3 Summary: The heuristics of creation

The "strategic triangle of business development" is a general attempt to identify a heuristic to place a business development process within a coherent and holistic context. A number of modifications have to be made to the original, more general model for it to be used specifically within the context of managing InnovationQuality. These have less impact on the basic components of the strategic triangle (current situation, framework conditions, objectives and strategy), but are more important for the steps of the innovation process that are derived from these.

Phase 1	Decision to innovate and develop InnovationQuality. A company must make a conscious decision to innovate and to lay down generation InnovationQuality objectives for this, and set up innovation projects.
Phase 2a	Analysis of the current situation affecting the company/project. The key questions this entails: > What value is currently being created in the areas where the innovation should be realized? And what is the scale of this value creation in relation to the overall business? > What is the current situation regarding the innovative capability of employees and managers enabling innovation, the focus on innovation in processes and structures, and partners and resources that foster innovation?
Phase 2b	Analysis of the framework conditions of the company/project. The key questions this entails: > What are the current and future competitive forces affecting the company/project as well as the political, macro-economic, social, technological, legal and environmental conditions? > What are the current megatrends and which future customer requirements can be anticipated as a result?
Phase 3	Definition and evaluation of the opportunities and threats for the company/project based on results of the analysis

>>

Phase 4	Definition of the objectives of an innovation project. An innovation project objective covers the following aspects: > Content of the outcome > Nature of the outcome (InnovationQuality in a narrower sense) > Purpose of the outcome > Timing of the outcome > Location of the outcome > Enablers of the outcome
Phase 5 and 6	Strategy definition, i.e., the planned approach to achieve business or innovation project objectives and implement the strategy. Naturally, the strategies and implementation of an innovation project differ markedly in terms of content, depending on the type of innovation (e.g., new product, new sales market, new business model, etc.). Nevertheless, on a formal level all innovation projects adhere to a similar logic: idea generation, conversion and diffusion.
Phase 7	Monitoring of achievement of InnovationQuality, i.e.: Were the things that were specifically laid down within the innovation objectives and the InnovationQuality objectives in particular actually achieved? If not, where are the deviations, deficits, or both? In principle, one could consider this step the "moment of truth" for InnovationQuality, since it is here that it becomes obvious whether and to what extent an innovation is actually of value. As a starting point for assessing this InnovationQuality the InnovationQuality formula can be used.
Phase 8	Entering back into the process.

Table 6: The eights stages of InnovationQuality development

As already highlighted at the beginning of this chapter, one of the weaknesses with existing innovation management models is that they are often based on a narrow interpretation of innovation, especially with product innovation. One noteworthy exception of this is the Innovation Helix model of Zillner and Krusche (2012). The Innovation Helix involves three steps, each divided into individual tasks, with specific tools for each task.

- Phase 1: Exploring, to identify relevant disruptions in the normal state of affairs (trends, inventions, etc.), and develop and prioritize new areas of innovation. The individual steps of this phase are: 1. Scouting, 2. Strategic analysis and 3. Strategic Operationalization.
- Phase 2: Designing, to give shape to new ideas and inventions. The individual steps of this phase are: 1. Need finding (research), 2. problem definition, 3. Ideation (developing concepts) and 4. Prototyping.
- Phase 3: Embedding, whereby initiatives that took place during previous phases are integrated into the daily business. The individual steps of this phase are: 1. Implementation, 2. Monitoring and 3. Evaluation.

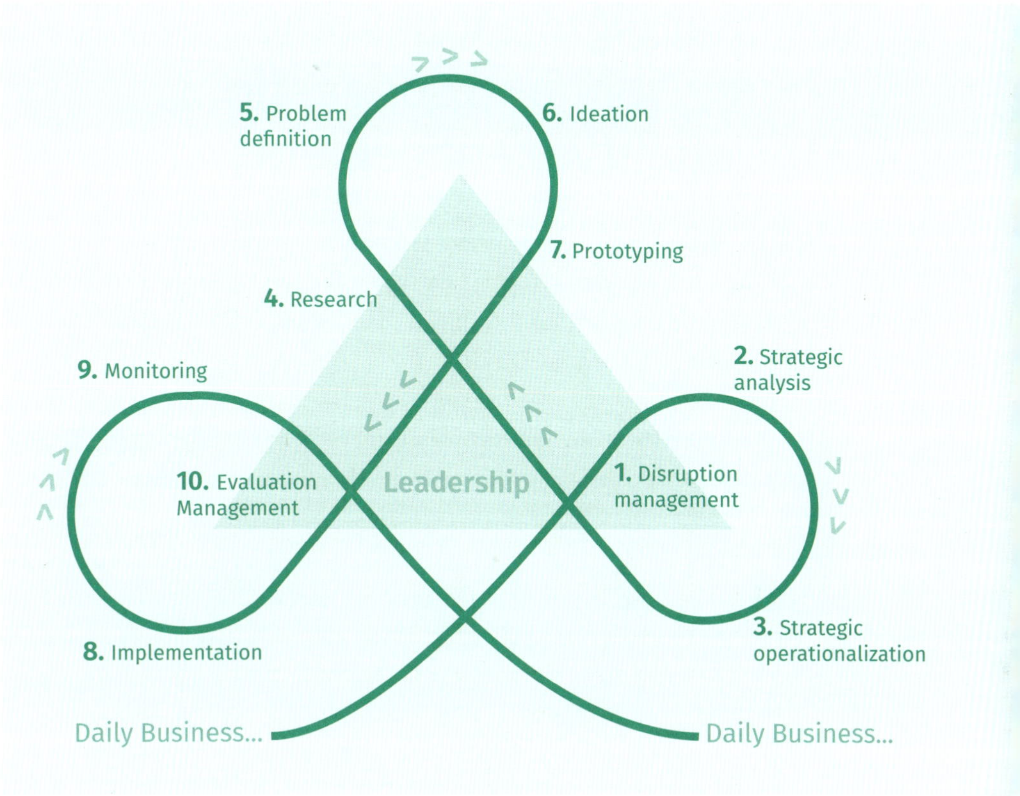

Figure 44: The Innovation Helix (Zillner and Krusche 2012)

The Innovation Helix has many things in common with the strategic triangle, not only in terms of its fundamental elements but also with respect to the steps it involves. As a result, both models should not be considered to compete with one another, but instead to complement one another, with possible weaknesses and blind spots of one being compensated for by the other. A key overlap between both models is that although innovation should be part of the everyday activities of a company, it should not be "daily business" or become "business as usual." Developing InnovationQuality should be an intrinsic part of how every company defines itself, manifested not only within tangible objectives and aims but also within the company culture. But innovation itself must be carried out to a large extent over and beyond operative activities. Innovation – assuming it should be of a high standard – thrives most when it is more or less distanced from the daily business and takes the form of innovation projects.

4 A final digression: Some homework and suggestions for quality management

The aim in the following is to examine a number of quality management concepts which could have a strong bearing on the term "InnovationQuality." To do this, some peculiarities are shown that are important to these concepts with respect to "InnovationQuality."

FMEA: Failure Mode and Effects Analysis makes it possible to identify potential weaknesses, errors and undesirable developments as soon as possible.

> Peculiarities with respect to InnovationQuality: Innovation projects are special because, in principle, the aim is to start with an idea and transform this into a successful application. FMEA methods play a particularly important role in this since they are based on the premise that there may be an error or a faulty process, and they attempt to evaluate factors relating to an error's probability of occurrence, identification and significance. One would be inclined to create a definition called "InFMEA," in other words "InnovationFMEA." To do this, one possibility would be to develop a method especially for the innovation process. A key success factor for such a method would be to systematically and methodically integrate experienced-based knowledge.

Perceived Quality: The concept of perceived quality is that the quality of an object stems from subjective appraisal (especially by the customer).

> Peculiarities with respect to InnovationQuality: The aim of an innovation is to achieve market success and this always goes hand in hand with successful market acceptance. Against this backdrop, the concept of perceived quality is linked to three possible consequences: 1. The most important condition is to know the market and especially to know the customer and, if possible, to also think further than the market and the customer. 2. What matters is keeping the customer on one's side by offering better and better InnovationQuality; in the long term, for something to be perceived as particularly innovative, "more of the same" is not sufficient – even if it is quality of a highly objective nature. 3. Aside from the objective quality

of products and services, subjective factors could also contribute to enhancements in InnovationQuality.

QFD, House of Quality: Quality Function Deployment is a methodical approach aimed at striking the right balance between customer desires and the provision of products and services. This method can be integrated into the more comprehensive concept of House of Quality.

> Peculiarities with respect to InnovationQuality: Both concepts can, on one hand, be used to systematically evaluate the interdependencies between desires and reality; for example, the design of a car is subject to constraints created by standards, statutory requirements, etc. On the other hand, these concepts provide a basis for quality planning when introducing features in keeping with market requirements. QFD and House of Quality could thus both make it possible to, at the very least, systematically edge closer to market-relevant features that could be conducive to the New becoming established.

PDCA: The Deming Cycle, also called the PDCA Cycle, is an intrinsic part of quality management. The fact that it is a cycle is a particular reflection of the desire to achieve a continuous improvement process (CIP). Using Plan, Do, Check and Act ensures that key performance indicators (KPIs) are used rigorously to check the success of each sub-process.

> Peculiarities with respect to InnovationQuality: The PDCA method is consciously intended to become a rigorously adhered to, fixed element at all stages of the innovation process. Accordingly, aside from installing creative and interactive procedures related to innovation, using the PDCA method constitutes an attempt to introduce a rational and rule-based counterbalance.

Cost impact monitoring throughout the entire life cycle: From the initial concept to the disposal of a product, the ending of a process or the completion of a service, it has to be ensured that the cost implications are understood as early as possible. At each stage, the cost impact of an error basically rises by a factor of 10. As a result, the consequences should be systematically captured early on to avoid fault-related costs at later stages.

> Peculiarities with respect to InnovationQuality: Procedures and methods have to be laid down within innovation management to act as an interface. With this approach, the aim is also to introduce a rational and rule-based counterbalance, over and above creative and interactive procedures related to innovation.

Strategic-Barriers: Distinctions are made between the following barriers

- Vision barriers
- People barriers
- Management barriers
- Acceptance barriers

> Peculiarities with respect to InnovationQuality: The processes often overlook the effect barriers have. Especially in the context of innovation management, consideration should be given to how they might be useful or how their negative impacts can be avoided.

Price/cost <-> value: With many development projects, it becomes clear that the terms price/cost and value are often confused with one another or used incorrectly. It is suggested that people are more likely to make a quick decision if they have an overview of costs and something is plausible.

> Peculiarities with respect to InnovationQuality: Care should be taken throughout the entire innovation process that costs are kept in proportion to value. Value or "being of value" is defined by customers and market needs. Being conscious of this is a necessary prerequisite for realizing an innovation in the first place. At each stage, it should be questioned whether something is of value. For example, this can be done by also using QFD in the House of Quality model.

SMART objectives: When setting the objectives for an innovation, Specific, Measurable, Attainable, Relevant and Timed should be inherent features for consideration.

> Peculiarities with respect to InnovationQuality: SMART objectives should be defined within the innovation process (see also the "Objectives" subchapter in the section on the "Strategic triangle"). Only then are the prerequisites fulfilled for systematic implementation.

Focusing on employee ability, motivation and permission: To integrate employees into processes as required, they must be in a position and have the skills to do this. Skills should be raised to a high standard with in-house training and targeted continuous professional development programs. Employees should be kept motivated so that they want to take on these tasks and if possible even take on the tasks with total enthusiasm. Ultimately, managers have to create the right atmosphere and establish the right conditions for this.

> Peculiarities with respect to InnovationQuality: Employees involved in the innovations process are under particular pressure. They have to be creative, master processes and pursue objectives rigorously. As a result, great value is attached to personal responsibility and task delegation. Employees have to be prepared for these tasks. Their ability has to match tasks and they have to be keen to take on tasks. Their managers have to express this "permission" aspect clearly and reinforce it.

Poka Yoke: Things being in the wrong place or badly installed can have a devastating impact on many processes. It is therefore important to try to avoid such "unfortunate errors." A tried and tested method for doing this is Poka-Yoke. Central to Poka-Yoke is the idea that people involved in planning consciously accept that mistakes could be made, e.g., in assembly, and that this will be reflected in negative results when something is completed.

> Peculiarities with respect to InnovationQuality: Networking plays an increasingly important role in innovation processes. Naturally, as a result there are a correspondingly large number of interfaces. These complex relationships between suppliers, manufacturers, clients and others mean that mistakes are practically pre-programmed. This is where it is important to use the Poka-Yoke method.

Conclusion

1 The imperative of the "Schumpeterian Entrepreneur"

Without wishing to exaggerate, we are witnesses and contemporaries of a time in which crises and haphazard events have simultaneously become a normality, never before witnessed in such short cycles. There are hardly any areas or any parts of the Earth that have been left unscathed by crises in recent years. The mother of all social crisis is, however, the inauspicious state or permanence that we have been witnessing since the economic turmoil of the late 2000s. What can be done in times when the global economy is under continual bombardment by crises? Without a doubt, in times of crisis a key task of entrepreneurship is to locate risk to the business and reduce it. However, as Roman Herzog, the former President of the Federal Republic of Germany once said: "Being a dynamic – or to a certain extent Schumpeterian – entrepreneur is and will remain a categorical imperative, the primary entrepreneurial responsibility and duty [...]: Whether in a small, medium-sized or large company, whether as the owner or employed as a manager." (translation of speech, Roman Herzog 1996). So what should companies do during and beyond moments of crisis to identify opportunities and, by realizing these opportunities, safeguard and extend their competitiveness in the long term?

In times of crisis there is a knee-jerk reaction and all entrepreneurial activity revolves around preserving the system, and sometimes that is the only remaining activity; use of the system is turned down a notch and sometimes system renewal is completely deactivated. At first glance, this reaction appears to be totally reasonable: Drawing a comparison with the Maslow hierarchy of needs, the main priority is to satisfy basic needs, which includes the preservation and safety of the system. It is only once this need has been satisfied that a system should think about other needs, or in simple terms: develop itself. But is a system that only wants to survive ultimately also capable of survival? Is a system that reduces its "existence" solely to preservation and safety – that does not allow for further development – not also inevitably doomed to fail if everything around it is changing more and more quickly and abruptly?

> *"The fundamental alternative for man is the choice between 'life' and 'death'. [...] Life means growing, developing, reacting.*

[...] Many people never face up to the clear alternative between the values of life and those of death, and as a result, they live in neither worlds, or become zombies, whose body is alive and whose soul is dead. (Translation, Fromm 1966: 192)

A system that wants to remain intact has to do everything the whole time. At the same time it has to: preserve the system (to keep going in the here and now); use the system (to "live"); renew the system (to safeguard and extend its future existence in the long term). What this means is that a business that concentrates solely on risk management during a crisis runs the risk of "extinction" if its economic environment undergoes change – change for the positive or the negative! It is especially during permanent, fatal, crisis-ridden phases that a system should be verging on a continuous state of chaos, preparing to completely change itself. Especially in times of sinister equilibrium, as is the case with today's economic crises, companies should bank more on system renewal and realizing the opportunities that emerge from a crisis – through innovation. Economic stimulus packages, government cash injections and other macro-political measures save businesses from financial fiascos. But it is only the New and the "creative destruction" of the old that goes hand in hand with the New – or to express this in one word: "innovation" – that is the driving force of sustainable economic development.

2 Innovation as the work of "creative people"

Innovation: Under no circumstances should this glamorous term be reduced in its meaning. Unfortunately, especially in an industrial nation like Germany, one often witnesses innovation almost dogmatically being placed on a par with "new products." We emphasize once again: Innovation should not be reduced to this level. Without a doubt, new products provide extremely important new impetus for safeguarding and developing competitiveness. But, for example, merely restricting oneself to launching increasingly superior and increasingly novel products is not enough to exploit the full market potential of such products. Because of this, we would like to leave you with a thought based on our own experience, a special thought that is particularly important for a business location like Germany: Current debate on the topic of "education" continually comes back to the close link between this issue and the competitiveness of a nation or organizations. Yet the discussion is restricted to MINT subjects, and science and technology education. This reduction of scope is problematical for two reasons:

1. Some people act as if potential insights or the transfer of knowledge from other disciplines – or drawing on the tremendous knowledge and ability of graduates in other areas – has no role to play in safeguarding and shaping the viability (or survival) of economies and businesses. Without a doubt, new products provide an important fillip when safeguarding and developing competitiveness. But over and beyond this, there is also a need to enter new supply and sales markets, to redesign organizations, and to establish new business models.

2. Some people act as if technological innovations only arise because the population has access to good scientific and technological education. Again, to state this clearly: Innovation happens when an idea is realized and to do this, it is not enough to be versed in technology or science; indeed it is not enough to supplement this specialist knowledge with a smattering of business administration know-how. To realize an idea requires a fundamental Schumpeterian attitude, so it requires the knowledge, ability and motivation to seize that advantage during change.

Innovation is an active deed of the kinds of people who have and are a "Schumpeterian personality." (cf. also Faix, Mergenthaler 2014) By Schumpeterian personality, Faix and Mergenthaler mean people:

> who draw on their broad and in-depth Bildung (in the German sense of self-cultivation) and their strong reasoning to think through the complex possible consequences of decisions and actions with circumspect and diligence

> who see and approach their Bildung as a person as the development of their fundamental being and existence, as a life-long challenge and liberty

> who have the knowledge and ability, but also the energy and courage to formulate and realize their own objectives in situations for which there is no template, nor standard, nor pre-formulated right or wrong

Creative personalities according to Faix and Mergenthaler are also marked out by the fact that they internalize and live what Kant described as a fundamental principle of ethics, as a "categorical imperative" of morality:

> the uncircumventable, unconditional feeling based on liberty – to have, to want, to realize something morally necessary and required, something "good" and something "just," with wisdom, courage and discretion

> the uncircumventable, unconditional feeling based on liberty – to have, to want, to never see or treat others or oneself as the means, but instead always as the end

After all, all people who are a creative personality are also – and particularly – marked by the fact that they are free and freedom-loving individuals in the most positive of senses. Creative personalities are not just value creators, they are also recalcitrant, they are not just brimming with talent but also contrariness. Intrinsic to their deeds are always values; reflected in their contemplation is always sustainability. They are individualists who use their knowledge, ability and motivation to make a vision of a "good life" possible for themselves and others.

Without people who think and act creatively, an idea remains merely a flight of fancy. It therefore requires innovators – people who make ideas happen. But how can a society, or how can an individual ensure that innovative potential is generated and is retained? The answer that we and many others give is: Bildung in the German sense, more Bildung and more still!

3 … and still not finished: Stimulus for more thought on the value of the New

Our book has provided a stimulus in a variety of ways – or at least that was our intention. On one hand, we hope it will make a fruitful contribution to the general debate surrounding the phenomenon of "innovation." On the other, we of course primarily deem it an attempt to explain and define the term "InnovationQuality." Our work can and must not be interpreted as the be-all and end-all. In keeping with scientific theory, we see this work as a paradigmatic foundation, a starting point, substance for further (and importantly more in-depth) examination of a number of minor and major questions and issues surrounding the topic of "InnovationQuality," which we could only touch on and only wanted to touch on this book. Possible questions and subjects include:

> What other kinds of innovation are conceivable and make sense? Our typology reflects all the types of areas a business could put thought into if it wanted to innovate. This typology of innovation thus provides a basis for all types of innovation activities in a business. Changes or additions to the existing types or even completely new types of innovation are thus to a certain extent "fundamental innovations," which change our concept of what an innovation is. Such innovation of an innovation opens up entirely new avenues to become more future-proof and more competitive.

> Which success factors or design principles are particularly important or fruitful for innovations of a greater value, given a specific social context? As already examined in the chapter on "A Model for InnovationQuality," we doubt whether this question can be answered clearly in a generally applicable way. So we hope there are reports that build on our model for InnovationQuality, particularly of a qualitative nature, and that they show what worked in a certain context and how – or perhaps what did not work. As discussed, such case reports will never put us in a position to entirely understand the phenomenon of "innovation," let alone control it. But every piece of input does at least shed light on the black box of "innovation."

> Which methods can be used to manage InnovationQuality? For example, what ways are there to analyze the current situation of a business in

terms of its enablers of innovation? How can megatrends be identified and evaluated? What is the best way to set innovation objectives?

> Where and how has the formula for InnovationQuality been used with success or what changes/additions are needed and where? What other formulae can be derived from the general InnovationQuality formula?

When we first started telling others about our plan to define and explain the term "InnovationQuality," we were given plenty of food for thought. Others had already had the courage to make a similar attempt – but when they realized what they had gotten themselves into, they decided they would rather leave it be. Something that actually spurred us on from the beginning was apparently something that made others give up: In their very essence, the terms "Innovation" and "Quality" could not be more different; on a superficial level, they are totally contrary to any kind of combination, they simply do not work well together. But it was exactly this predicament that motivated us, true to a principle: the greater the suspense, the greater the outcome. Whether and to what extent our thinking will be effective in the long term, or even beneficial, is of course something we cannot tell. Ultimately, one thing the reader might grant us is that we were willing to take the risk of "going down in style." But this will only be the case if there is no discussion or if discussion does not at least result in something incrementally new. And we would like to drive this process and will make the twin/coupled phenomenon of "InnovationQuality" the subject matter of events suited to the topic. This will be supported by the following (hopefully changing) microsite:

www.steinbeis.de/inq

Our aim is to expand on certain aspects relating to this topic with some of our own articles and, importantly, articles submitted by you – as additions, to point things out as examples of use, with the results of events, etc. In this, too, something one has to grant us is that we are doing all of this despite resistance – and we will be curious to see what it leads to. One thing that certainly will not happen is nothing at all – the work between the authors and the discussions we have had, as well as the "working groups" and things we have undertaken ourselves as outlined earlier, have borne fruit – with the prospect of more to come.

List of references

Adams, R.; Bessant, J.; Phelps, R. (2006): Innovation management measurement: a review. In: International Journal of Management Reviews, Vol. 8 No. 1, p. 21–47.

Baregheh, A.; Rowley, J.; Sambrook, S. (2009); Towards a multidisciplinary definition of innovation. In: Management Decision, Vol. 47, No. 8, p. 1323–1339.

BDI & Deutsche Telekom Stiftung (BDI&DTS) (2011): Innovationsindikator 2011. www.innovationsindikator.de/fileadmin/user_upload/Dokumente/Innovationsindikator_2011.pdf; last checked on 11.2.2013.

Beck, U. (1997): Was ist Globalisierung? Frankfurt. [Beck, U. (2000): What is Globalization? Cambridge.]

Bessant, J.; Lamming, R.; Noke, H.; Phillips, W. (2005): Managing innovation beyond the steady state. In: Technovation, Vol. 25 No. 12, p. 1366–1376.

Bleicher, K. (2004): Das Konzept Integriertes Management, 7. Auflage. Frankfurt am Main u. a.

Bormannn, I. (2012): Indikatoren für Innovation – ein Paradox? In: Bormann, I.; Aderhold, J.; John, R. (Hrsg.): Indikatoren des Neuen. Innovation als Sozialmethodologie oder Sozialtechnologie? Wiesbaden, p. 39–55.

Bruhn, M. (2013): Qualitätsmanagement für Dienstleistungen. Handbuch für ein erfolgreiches Qualitätsmanagement. Grundlagen – Konzepte – Methoden, 9., vollständig überarbeitete und erweiterte Auflage. Heidelberg, Berlin.

Capra, F. (1992): Wendezeit, 2. Auflage. München. [Capra, F. (1982): The Turning Point. London.]

Collins, J. (2001): Good to Great. Why some companies make the leap and others don't... New York.

Damanpour, F.; Schneider, M. (2006): Phases of the adoption of innovation in organizations: effects of environment, organization and top managers. In: British Journal of Management, Vol. 17 No. 3, p. 215–36.

Darwin, C. (1859/1909): On the origin of species by means of natural selection, or the preservation of favoured races in the struggle for life. New York.

Department of Trade of the UK (2003): Innovation Report. Department of Trade. London.

Drucker, P. F. (1981): Innovationsmanagement für Wirtschaft und Politik. Düsseldorf.

Dutta, S. (Hrsg.) (2011a): The Global Innovation Index 2011. http://www.globalinnovationindex.org/gii/GII%20COMPLETE_PRINTWEB.pdf, last checked on 11.2.2013.

Dutta, S. (Hrsg.) (2011b): The Global Innovation Index 2011. Executive Summary www.globalinnovationindex.org/gii/GII%202011%20Executive%20Summary.pdf, last checked on 11.2.2013.

Ettlie, J. E.; Bridges, W. P.; O'Keefe, R. D. (1984): Organizational strategy and structural differences for radical versus incremental innovation. In: Management Science, Vol. 30 No. 6, p. 682–695.

Faix, W. G. (2008): Die Unternehmensentwicklung zu Wachstum und Globalisierung. In: Faix, W. G.; Keck, G.; Kisgen, S.; Mezger, P.; Sailer, J.; Schulten, A. (Hrsg.): Management von Wachstum und Globalisierung. Best Practice Band 3. Stuttgart, p. 17–71.

Faix, W. G.; Buchwald, C.; Wetzler, R. (1991): Skill Management. Qualifikationsplanung für Unternehmen und Mitarbeiter. Wiesbaden.

Faix, W. G., Buchwald, C.; Wetzler, R. (1994): Der Weg zum schlanken Unternehmen. Landsberg.

Faix, W. G.; Mergenthaler, J. (2014): Die schöpferische Kraft der Bildung. Über Innovation, Unternehmertum, Persönlichkeit und Bildung, 2. Auflage. Stuttgart.

Freund, D. (2013): Wertschöpfende und innovationsorientierte Unternehmensführung. Berlin, Heidelberg.

Gillwald, K. (2000): Konzepte sozialer Innovation. Wissenschaftszentrum, Berlin. http://bibliothek.wzb.eu/pdf/2000/p00-519.pdf

Gronemeyer, M. (2000): Immer wieder neu oder ewig das Gleiche. Darmstadt.

Hansen, M. T.; Birkinshaw, J. (2012): The Innovation Value Chain. In: Harvard Business Review (June), p. 121–130.

IBM (2008): IBM Global CEO Study. The Enterprise of the Future. www.ibm.com/services/de/bcs/html/ceostudy.html

IBM (2010): IBM Global CEO Study. Capitalizing on Complexity. http://www-935.ibm.com/services/de/ceo/ ceostudy2010/

IBM (2012): IBM Global CEO Study. Leading through Connections. http://www-935.ibm.com/services/de/ceo/ceostudy.

Kamiske, G. F. (Hrsg.) (1996): Rentabel durch TQM, Return on Quality – ROQ. Berlin.

Lachenmaier, S.; Woessmann, L. (2004): Does Innovation cause Exports? Evidence from exogenous Innovation impulses and obstacles using german micro data. CESIFO Working Papers No. 1178 Category 7: Trade Policiy. http://ideas.repec.org/a/oup/oxecpp/v58y2006i2p317-350.html

Lachenmaier, S.; Rottmann, H. (2007a): Effects of Innovation on Employment: A dynamic panel analysis. CESIFO Working Papers No. 2015 Category 9: Industrial Organisation. http://ideas.repec.org/p/mse/cesdoc/r07036.html

Lachenmaier, S.; Rottmann, H. (2007b): Employment Effects of Innovation at the Firm Level. Ifo Working Paper No. 27. http://ideas.repec.org/p/ces/ifowps/_27.html

Lachenmaier, S.; Woessmann, L. (2004): Does Innovation cause Exports? Evidence from exogenous Innovation impulses and obstacles using german micro data. CESIFO Working Papers No. 1178 Category 7: Trade Policiy. http://ideas.repec.org/a/oup/oxecpp/v58y2006i2p317-350.html

Lachenmaier, S.; Rottmann, H. (2007a): Effects of Innovation on Employment: A dynamic panel analysis. CESIFO Working Papers No. 2015 Category 9: Industrial Organisation. http://ideas.repec.org/p/mse/cesdoc/r07036.html

Lachenmaier, S.; Rottmann, H. (2007b): Employment Effects of Innovation at the Firm Level. Ifo Working Paper No. 27. http://ideas.repec.org/p/ces/ifowps/_27.html

List, F. (1930/1841): Das nationale System der politischen Ökonomie. 5. Auflage. Jena. [List, F. (1974/1885): The National System of Political Economy. London, New York.]

Löhn, J. (1995): LöhnMethode – Ziele. Waldkirch.

Luhmann, N. (1988): Erkenntnis als Konstruktion. Bern.

Luhmann, N. (1997): Die Gesellschaft der Gesellschaften. Frankfurt a.M. [Luhmann, N. (1997): Theory of Society, Stanford.]

Mann, R. (1990): Das ganzheitliche Unternehmen. Bern, München, Wien

Maslow, A. (1954): Motivation und personality. New York.

Matzler, K.; von den Eichen, S. (2012): Innovators Dilemma. Warum etablierte Unternehmen bei bahnbrechenden Innovationen scheitern. In: Peter Granig (Hrsg.): Die Kunst der Innovation. Von der Idee zum Erfolg. Wiesbaden, p. 51–62.

Mergenthaler, J.; Faix, W. G. (2014): Steinbeis-Innovationsstudie. Eine Metastudie über die Innovationsfähigkeit und -tätigkeit der Volkswirtschaften von Brasilien, China, Deutschland, der Schweiz und der USA. Stuttgart.

Merton, R. K. (1968): The Matthew Effect in Science. The Reward and Communication Systems of Science are Considered. In: Science, 159, p. 56–63.

Nagel, K. (1995): Die 6 Erfolgsfaktoren des Unternehmens. Landsberg.

OECD (2005): The Measurement of Scientific and Technological Activities. Oslo Manual. Guidelines for Collecting and Interpreting Innovation Data, 3rd Edition. Paris.

Oelsnitz, D. von; Stein, V.; Hahmann, M. (2007): Der Talente-Krieg. Personalstrategie und Bildung im globalen Kampf um Hochqualifizierte. Bern u. a.

Price, D. J. (1971): Some Remarks on Elitism in Information and the Invisible College Phenomenon in Science. In: Journal of the American Society for Information Science, March-April 1971, p. 74–75.

Raisch, S.; Probst, G.; Gomez, P. (2007): Wege zum Wachstum. Wie Sie nachhaltigen Unternehmenserfolg erzielen. Wiesbaden.

Rasner, C.; Füser, K.; Faix, W. G. (1999): Das Existenzgründer-Buch, 4. Auflage. Landsberg.

Revans, R. W. (1983): The ABC of Action Learning. Bromley.

Rogers, E. M. (2003): Diffusion of Innovations. 5th Edition. New York.

Rosa, H. (2008): Im Wirbel der Beschleunigungsspirale. In: Spektrum der Wissenschaft Februarausgabe, p. 82–87.

Rust, R. T.; Zahorik, A. J.; Keiningham, T. L. (1994): Return on Quality (ROQ): Making Service Quality Financially Accountable. In: Journal of Marketing Vol. 59, p. 58–70.

Scanlon, J. (2006): How to turn money into innovation. In: Business Week, 14. 11.2006.

Schuhmacher, E. F. (1977): Die Rückkehr zum menschlichen Maß (Small is Beautifull). Reinbeck. [Schumacher, E.F. (1973): Small is Beautiful: (A Study of) Economics as if People Mattered. London.]

Schumpeter (1942/1975): Capitalism, Socialism and Democracy. New York.

Schumpeter, J. A. (1947): The Creative Response in Economic History. The Journal of Economic History 7 (2), p. 149–159.

Schumpeter, J.A. (1934): The Theory of Economic Development. Cambridge.

Siemens AG (1994): Zukunft braucht Herkunft. München.

Smolny, W.; Schneeweis, T. (1999): Innovation, Wachstum und Beschäftigung. Eine empirische Untersuchung auf der Basis des ifo Unternehmenspanels. In: Jahrbücher für Nationalökonomie und Statistik 218, Heft 3+4, p. 457–472.

World Economic Forum (WEF) (Hrsg.) (2011): The Global Competitiveness Report 2011-2012. www.weforum.org/pdf.php?download=103512, last checked on 11.2.2013.

Weingart, R. (2011): Wissenschaftspolitik als Innovationspolitik: Anspruch und Wirklichkeit. In: Gegenworte, 26. Heft Herbst 2011, p. 18–23.

Wiebe, F. (2012): Über das Risiko, zu viele Risiken zu kennen. In: Handelsblatt vom Mittwoch 9.8.2012, Nr. 152, p. 10.

Zahra, S. A.; Covin, J.G. (1994): The financial implications of fit between competitive strategy and innovation types and sources. In: The Journal of High Technology Management Research, Vol. 5 No. 2, p. 183–211.

Zillner, S.; Krusche, B. (2012): Systemisches Innovationsmanagement. Grundlagen – Strategien – Instrumente. Stuttgart.